ASE Correlated Task Sheets

for

Advanced Engine Performance Diagnosis

EIGHTH Edition

Curt Ward

Pearson

Product Management: Derril Trakalo
Content Production: Isha Sachdeva

2 2026

Pearson

ISBN 10: 0-13-809653-8
ISBN 13: 978-0-13-809653-3

TABLE OF CONTENTS

Pearson

ASE Education Foundation

Vehicle Identification Number

Meets ASE Task: A3 – A-1 – P1

Evaluation (Enter number from 4, 3, 2, 1): ☐

Name: ☐ **Date:** ☐ **Time on Task:** ☐

Make/Model/Year: ☐ **VIN:** ☐

☐ **1.** The first number or letter designates the **country of origin =** ☐

1 = United States	6 = Australia	L = China	V = France
2 = Canada	8 = Argentina	R = Taiwan	W = Germany
3 = Mexico	9 = Brazil	S = England	X = Russia
4 = United States	J = Japan	T = Czechoslovakia	Y = Sweden
5 = United States	K = Korea	U = Romania	Z = Italy

☐ The model of the vehicle is commonly the fourth or fifth character. **Model?** ☐

☐ The eighth character is often the engine code. (Some engines cannot be determined by the VIN number.) **Engine code:** ☐

☐ The tenth character represents the year on all vehicles. See the following chart. ☐

VIN Year Chart (The pattern repeats every 30 years.) **Year?** ☐

A = 1980/2010	J = 1988/2018	T = 1996/2026	4 = 2004/2034
B = 1981/2011	K = 1989/2019	V = 1997/2027	5 = 2005/2035
C = 1982/2012	L = 1990/2020	W = 1998/2028	6 = 2006/2036
D = 1983/2013	M = 1991/2021	X = 1999/2029	7 = 2007/2037
E = 1984/2014	N = 1992/2022	Y = 2000/2030	8 = 2008/2038
F = 1985/2015	P = 1993/2023	1 = 2001/2031	9 = 2009/2039
G = 1986/2016	R = 1994/2024	2 = 2002/2032	
H = 1987/2017	S = 1995/2025	3 = 2003/2033	

Pearson

ASE Education Foundation

Vehicle Service History

Meets ASE Task: (A1 through A8-A-2) P-1

Evaluation (Enter number from 4, 3, 2, 1): ☐

Name: ☐ **Date:** ☐ **Time on Task:** ☐

Make/Model/Year: ☐ **VIN:** ☐

☐ **1.** Search vehicle history (check all that apply).

- ☐ Computerized data base (electronic file if previous service work)
- ☐ Files (hard copy of previous service work)
- ☐ Customer information (verbal)
- ☐ Other (describe): ☐

☐ **2.** What electrical-related repairs have been performed in this vehicle?

☐ **3.** From the information obtained, has the vehicle been serviced regularly?

- ☐ ☐ Yes (describe the service intervals) ☐
- ☐ ☐ No (why?) ☐

☐ **4.** Based on the service history information, is the service record helpful? Why or why not?

Pearson

Education Foundation

Technical Service Bulletins

Meets ASE Task: (A1 through A8-A-2) P-1

Evaluation (Enter number from 4, 3, 2, 1): ☐

Name: ☐ **Date:** ☐ **Time on Task:** ☐

Make/Model/Year: ☐ **VIN:** ☐

☐ **1.** Technical service bulletins can be accessed through (check all that apply):

- ☐ Internet site(s), specify ☐
- ☐ Paper bulletins, specify source ☐
- ☐ CD ROM bulletins, specify source ☐
- ☐ Other (describe): ☐

☐ **2.** List all electrical-related technical service bulletins that pertain to the vehicle/engine being serviced.

Number	Description/Correction

☐ **3.** Based on this research, is the information located helpful?

☐ [ ] Yes, why? [ ]

☐ [ ] No, why not? [ ]

Pearson

Education Foundation

Vehicle Service Information

Meets ASE Task: A1 – A-1 – P1

Evaluation (Enter number from 4, 3, 2, 1): ☐

Name: ☐ **Date:** ☐ **Time on Task:** ☐

Make/Model/Year: ☐ **VIN:** ☐

☐ **1.** Check the service information for safety precautions and technical service bulletins. (describe):

☐ **2.** Review the service history for outstanding recalls. (describe):

☐ **3.** Check the service information for any special tools needed to service the engine. (describe):

☐ **4.** Type and quantity of engine oil used (describe):

☐ **5.** Type and quantity of engine coolant used (describe):

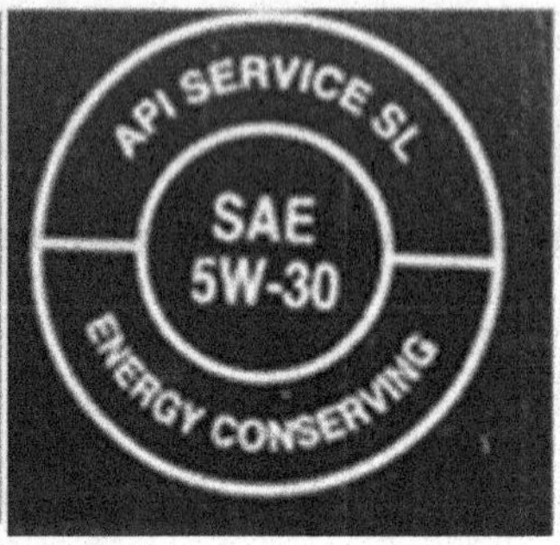

Pearson

Education Foundation

Vehicle Safety Certification Label

Meets ASE Task: (A1 through A8-A-2) P-1

Evaluation (Enter number from 4, 3, 2, 1): ☐

Name: ☐ **Date:** ☐ **Time on Task:** ☐

Make/Model/Year: ☐ **VIN:** ☐

☐ **1.** Describe the location of the Vehicle Safety Certification Label (usually located on the driver's side pillar post).

☐ **2.** What is the month and year the vehicle was manufactured?

Month = ☐

Year = ☐

☐ **3.** What is the gross vehicle weight rating (GVWR)?

☐ **4.** What is the gross axle weight rating (GAWR)?

☐ **5.** What is the vehicle type?

☐ Passenger Car ☐ Light duty truck ☐ Incomplete vehicle

Pearson

Education Foundation

Work Order

Meets ASE Task: A1 thru A8 – A-1 - P-1

Evaluation (Enter number from 4, 3, 2, 1): ☐

Name: ______ **Date:** ______ **Time on Task:** ______

Make/Model/Year: ______ **VIN:** ______

☐ **1.** List the items about the **vehicle** that should be included on the work order (also called a repair order - R.O).

a. ______ e. ______

b. ______ f. ______

c. ______ g. ______

d. ______ h. ______

☐ **2.** List the information about the **driver/owner** that should be included on the work order.

a. ______

b. ______

c. ______

d. ______

3. List the three Cs (concern, cause, and correction) that the service technician should write on the work order for a repair that includes a diagnosis of the problem (concern), the replacement part, and the verification of repair.

a.

b.

c.

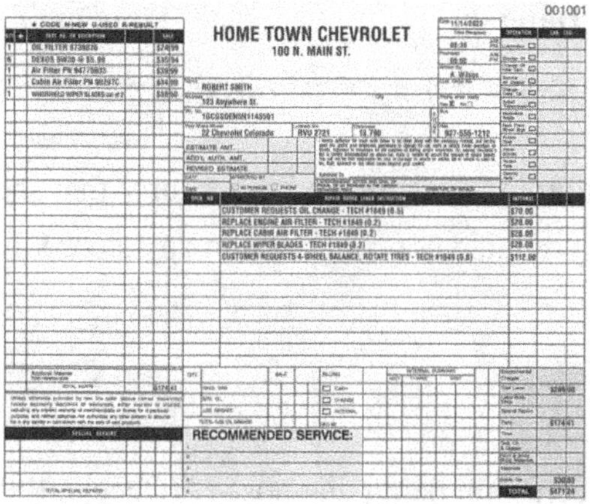
001001

HOME TOWN CHEVROLET
100 N. MAIN ST.

QTY	PART NO. OR DESCRIPTION	SALE
1	OIL FILTER	[illegible]
6	DEXOS 5W30 @ $5.99	[illegible]
1	Air Filter PN [illegible]	[illegible]
1	Cabin Air Filter PN [illegible]	[illegible]
1	WINDSHIELD WIPER BLADES	[illegible]

ROBERT SMITH
123 Anywhere St.
22 Chevrolet Colorado

ESTIMATE AMT.
ADD'L AUTH. AMT.
REVISED ESTIMATE

REPAIR ORDER LABOR INSTRUCTION	AMOUNT
CUSTOMER REQUESTS OIL CHANGE - TECH #1849 (0.5)	$70.00
REPLACE ENGINE AIR FILTER - TECH #1849 (0.2)	$28.00
REPLACE CABIN AIR FILTER - TECH #1849 (0.2)	$28.00
REPLACE WIPER BLADES - TECH #1849 (0.2)	$28.00
CUSTOMER REQUESTS 4-WHEEL BALANCE, ROTATE TIRES - TECH #1849 (0.8)	$112.00

RECOMMENDED SERVICE:

TOTAL

Pearson

ASE Education Foundation

No-Code Diagnosis of Driveability Concerns

Meets ASE Task: A8 – B-8 – P-1

Evaluation (Enter number from 4, 3, 2, 1): ☐

Name: ☐ **Date:** ☐ **Time on Task:** ☐

Make/Model/Year: ☐ **VIN:** ☐

☐ **1.** Check service information for the specified procedures that should be followed when diagnosing a driveability-related concern without any stored diagnostic trouble codes.

☐ **2.** Which driveability concerns are being diagnosed? (Check all that apply.)

- ☐ ☐ Hesitation during acceleration
- ☐ ☐ Rough or unstable idle
- ☐ ☐ Stalling
- ☐ ☐ Missing
- ☐ ☐ Other (describe) ☐

☐ **3.** Based on the results of the scan tool diagnosis, what is the necessary action?

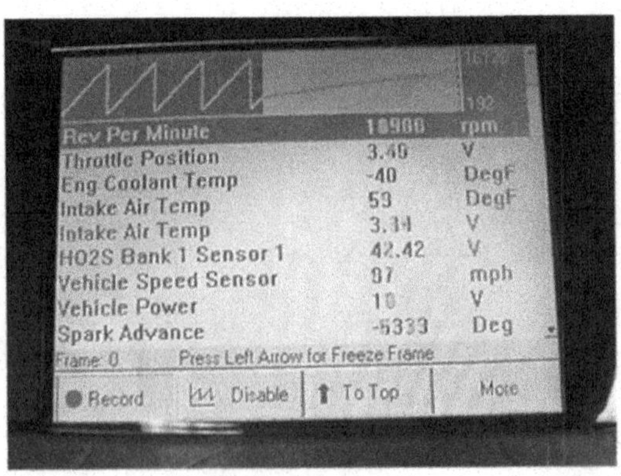

Pearson

Education Foundation

Flexible Fuel Vehicle Identification

Meets ASE Task: A8 – A-1 – P-1

Evaluation (Enter number from 4, 3, 2, 1): ☐

Name: ______ **Date:** ______ **Time on Task:** ______

Make/Model/Year: ______ **VIN:** ______

☐ **1.** Many vehicles are manufactured that are capable of operating on fuels other than, or in addition to, regular gasoline. Check service information and determine where it is stated on the vehicle if it can use a fuel other than gasoline and what fuel(s) it can use. Select one vehicle model from each major manufacturer.

Manufacturer	Model	Year	Location of Information	Fuel(s)
General Motors				
Chrysler				
Honda				
Toyota				
Nissan				
Hyundai				
Kia				
Other				
Other				

Pearson

Education Foundation

Alternative Fuel Power and Economy Specifications

Meets ASE Task: A8 – A-1 – P-1

Evaluation (Enter number from 4, 3, 2, 1): ☐

Name: ☐ **Date:** ☐ **Time on Task:** ☐

Make/Model/Year: ☐ **VIN:** ☐

☐ **1.** With vehicles that are designed to use gasoline and alternative fuel, such as E-85, manufacturers usually rate the horsepower, torque, and EPA fuel economy ratings separately. Select four flex-fuel vehicles and using service information locate the values for both gasoline and the alternative fuel.

Vehicle #1 Make ____________ **Model** ____________ **Year** ____ **Engine** __________

	HP	Torque	City MPG	Highway MPG
Gasoline				
Alternative Fuel				

Vehicle #2 Make ____________ **Model** ____________ **Year** ____ **Engine** __________

	HP	Torque	City MPG	Highway MPG
Gasoline				
Alternative Fuel				

Vehicle #3 Make ____________ **Model** ____________ **Year** ____ **Engine** __________

	HP	Torque	City MPG	Highway MPG
Gasoline				
Alternative Fuel				

Vehicle #4 Make ____________ **Model** ____________ **Year** ____ **Engine** __________

	HP	Torque	City MPG	Highway MPG
Gasoline				
Alternative Fuel				

Pearson

Education Foundation

Alcohol Content in Gasoline

Meets ASE Task: A8 – D-7 – P-1

Evaluation (Enter number from 4, 3, 2, 1): ☐

Name: ☐ **Date:** ☐ **Time on Task:** ☐

Make/Model/Year: ☐ **VIN:** ☐

☐ **1.** Pour suspect gasoline into a small clean beaker or glass container.

☐ **2.** Carefully fill the graduated cylinder to the 10-mL mark.

☐ **3.** Add 2 mL of water to the graduated cylinder by counting the number of drops from an eyedropper. (Before performing the test, the eyedropper must be calibrated to determine how many drops equal 2.0 mL.)

☐ **4.** Put the stopper in the cylinder and shake vigorously for 1 minute. Relieve built-up pressure by occasionally removing the stopper. Alcohol dissolves in water and will drop to the bottom of the cylinder.

☐ **5.** Place the cylinder on a flat surface and let it stand for 2 minutes.

☐ **6.** Take a reading near the bottom of the cylinder at the boundary between the two liquids.

☐ **7.** For percent of alcohol in gasoline, subtract 2 from the reading and multiply by 10.

The reading is 3.1 mL: 3.1 - 2 = 1.1 X 10 = 11% alcohol

The reading is 2.0 mL: 2 - 2 = 0 X 10 = 0% alcohol (no alcohol)

If the increase in volume is 0.2% or less, it may be assumed that the test gasoline contains no alcohol. Alcohol content can also be checked using an electronic tester.

☐ **8.** Based on the test results, what action is necessary?

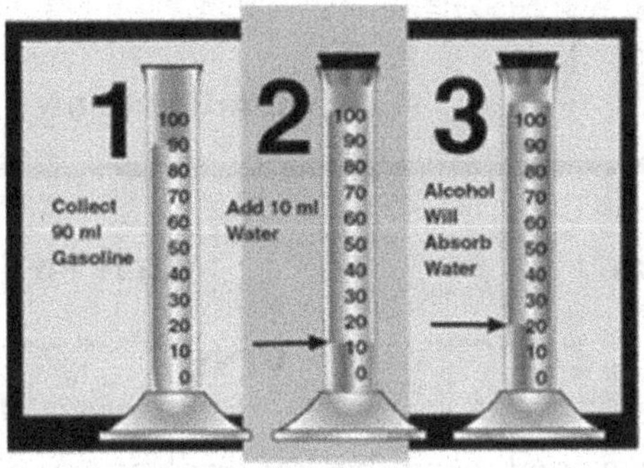

Pearson

Education Foundation

Digital Multimeter Use for Electrical Problems

Meets ASE Task: A6 – A-5 – P-1

Evaluation (Enter number from 4, 3, 2, 1): ☐

Name: ☐ **Date:** ☐ **Time on Task:** ☐

Make/Model/Year: ☐ **VIN:** ☐

☐ **1.** Check service information on the correct procedure for checking charging system voltage.

- ☐ Which position on the digital multimeter should be selected to check charging system voltage? ☐
- ☐ What is the specified voltage? ☐
- ☐ What is the actual voltage? ☐

☐ **2.** Check service information regarding the specified resistance for spark plug wires.

- ☐ Which position on the DMM should be selected to measure spark plug wire resistance? ☐
- ☐ What is the specified resistance? ☐
- ☐ What is the actual (measured) resistance? ☐

☐ **3.** Check service information regarding the specified battery drain test (parasitic draw) test procedures.

☐ Which position on the DMM should be selected to measure battery electrical drain?

☐ What is the specified maximum battery electrical drain?

☐ What is the actual (measured) battery electrical drain?

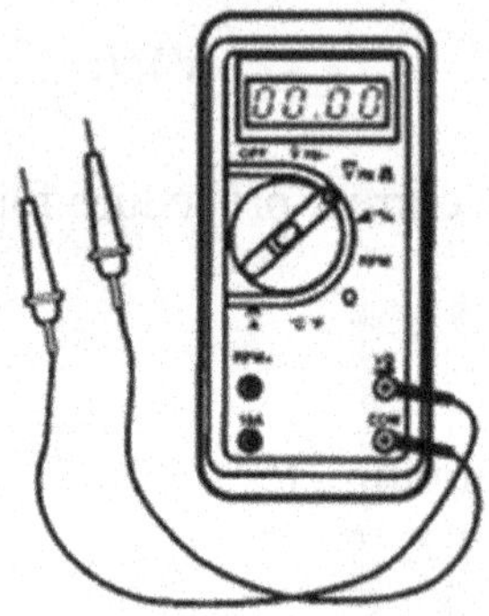

Pearson

Education Foundation

Test Light Usage

Meets ASE Task: A6 - A-7- P-1

Evaluation (Enter number from 4, 3, 2, 1): ☐

Name: ☐ **Date:** ☐ **Time on Task:** ☐

Make/Model/Year: ☐ **VIN:** ☐

☐ **1.** Check service information regarding which wire color(s) is the brake light and which is the tail light. Using a test light, check for voltage to the brake lights and tail lights at the rear of the vehicle.

☐ Brake light: ☐ OK ☐ NOT OK (describe fault) ☐

☐ Tail light: ☐ OK ☐ NOT OK (describe fault) ☐

☐ **2.** Using a test light, check all the fuses in the vehicle.

☐ **3.** Based on the above activities, what is the necessary action?

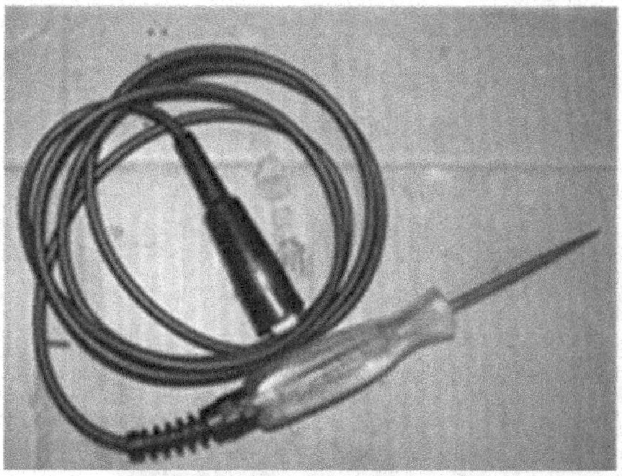

Pearson

ASE Education Foundation

Circuit Testing Using a Fused Jumper Wire

Meets ASE Task: A6 - A-8 – P-1

Evaluation (Enter number from 4, 3, 2, 1): ☐

Name: ☐ **Date:** ☐ **Time on Task:** ☐

Make/Model/Year: ☐ **VIN:** ☐

☐ **1.** Check service information for a diagnostic test procedure that includes the use of a fused jumper wire.

☐ ***CAUTION:*** A fused jumper wire should never be used to bypass an electrical load device. A fused jumper wire should only be used to bypass circuit control devices such as switches or relays.

☐ **2.** A horn circuit is a commonly used circuit to show the use of a fused jumper wire.

- ☐ Locate the horn (describe the location): ☐
- ☐ Disconnect the wire from the horn.
- ☐ Connect one end of the fused jumper wire to the terminal of the horn.
- ☐ Touch the other end of the fused jumper wire to the positive (+) terminal of the battery.
- ☐ The horn should work. OK ☐ ☐ NOT OK ☐ ☐

☐ **3.** Based on the test results, what is the necessary action?

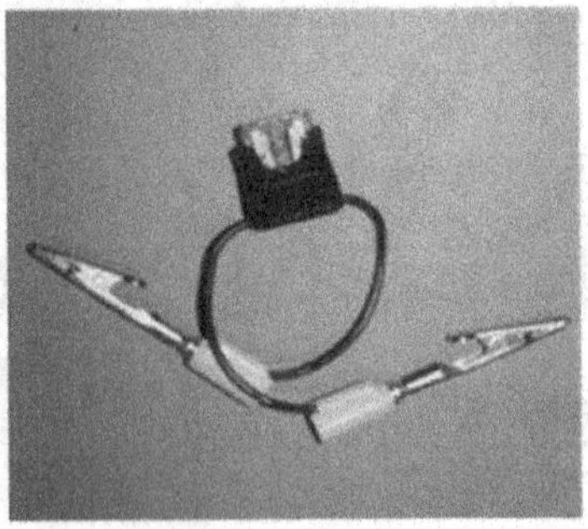

Pearson

Education Foundation

Oscilloscope Testing

Meets ASE Task: A6 – A-13 – P-1

Evaluation (Enter number from 4, 3, 2, 1): ☐

Name: ☐ **Date:** ☐ **Time on Task:** ☐

Make/Model/Year: ☐ **VIN:** ☐

☐ **1.** Check service information for the recommended usage of an oscilloscope to diagnose electrical or electronic concerns. Describe the tests recommended.

☐ **2.** According to service information, which system can/should be tested using an oscilloscope? (Check all that apply.)

- ☐ ☐ Wheel speed sensor
- ☐ ☐ Crankshaft position sensor
- ☐ ☐ Camshaft position sensor
- ☐ ☐ Secondary ignition
- ☐ ☐ Serial data communication circuits
- ☐ ☐ Motor control circuits
- ☐ ☐ Other (describe) ☐

☐ **3.** Which system(s) was tested using an oscilloscope?

☐ **4.** Based on the test results, what is the necessary action?

Pearson

ASE Education Foundation

Oscilloscope Setup and Adjustment

Meets ASE Task: A6 – A-13 – P-1

Evaluation (Enter number from 4, 3, 2, 1): ☐

Name: ☐ **Date:** ☐ **Time on Task:** ☐

Make/Model/Year: ☐ **VIN:** ☐

☐ **1.** After powering up the oscilloscope, set the time base to 20 milliseconds (ms) per division. Record the total time displayed on the screen.

☐ **2.** Adjust the volts per division to 1 volt per division. Record the total voltage displayed on the screen.

☐ **3.** Configure the scope to display two channels.

☐ **4.** Setup a trigger on channel A or 1. Adjust the trigger level to 2.8 volts with the slope in a positive direction.

☐ **5.** Connect a breakout box to the data link connector (DLC) and connect the scope leads to the communication circuit(s)

☐ **6.** With the ignition key in the on position, observe the pattern. (describe):

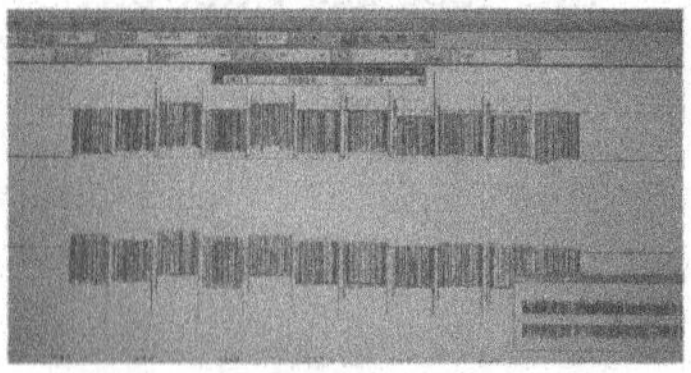

Pearson

ASE Education Foundation

Scan Tool Diagnosis

Meets ASE Task: A-8 – B-7 – P-1, A-8 – B-8 – P-1

Evaluation (Enter number from 4, 3, 2, 1): ☐

Name: ☐ **Date:** ☐ **Time on Task:** ☐

Make/Model/Year: ☐ **VIN:** ☐

☐ **1.** Connect a scan tool with the key on, engine off (KOEO), and disconnect the electrical connection from the throttle position (TP) sensor. Wait 3 seconds.

☐ **2.** Is the malfunction indicator lamp (MIL or check engine) on? ☐ Yes ☐ No

☐ **3.** Retrieve the Diagnostic Trouble Codes (DTCs) using the scan tool and record them below.

☐ **4.** Did the scan tool display both a generic OBD II (P0xxx) code *and* a manufacturer's specific DTC (P1xxx) code?

Yes ☐ No ☐

☐ **5.** Using a scan tool, view the freeze frame created when the DTC was set.

☐ OK (freeze frame was set) ☐ No (freeze frame was not set)

☐ **6.** Reconnect the electrical connection from the throttle position (TP) sensor.

☐ **7.** Clear the stored DTCs using the scan tool.

Pearson

Education Foundation

Replace Water Pump II

Meets ASE Task: A1 – D-8 – P-1

Evaluation (Enter number from 4, 3, 2, 1):

Name: **Date:** **Time on Task:**

Make/Model/Year: **VIN:**

☐ **1.** Check service information for the recommended procedures to inspect and replace the water pump (describe recommended procedures).

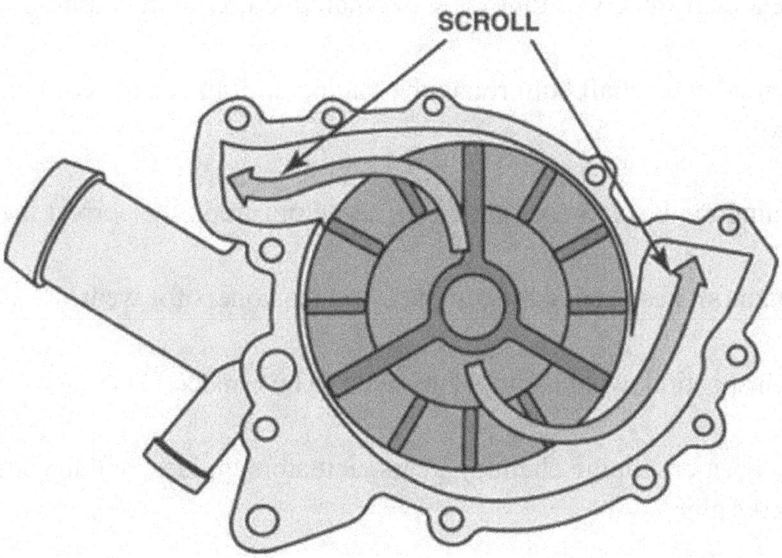

☐ **2.** What are the torque specifications for the water pump fasteners?

Pearson

Education Foundation

Inspect and Replace Timing Chain(s) and Gears

Meets ASE Task: A1 – B-6 – P-1

Evaluation (Enter number from 4, 3, 2, 1):

Name: **Date:** **Time on Task:**

Make/Model/Year: **VIN:**

☐ **1.** Check service information to determine the specified procedure and special tools needed to inspect and replace the timing chain(s), gears, actuators, guides, and tensioners. List the special tools needed:

☐ **2.** Remove the engine front cover following the manufacturer's procedures.

☐ **3.** Using the original crankshaft bolt, rotate the engine until the crankshaft gear is in the desired location.

☐ **4.** Verify the timing marks on the variable cam actuators are in the correct location.

☐ **5.** Inspect the chain(s), gears, actuators, guides, and tensioners for wear.

☐ **6.** Based on the inspection, what components need to be replaced?

☐ **7.** Install the replacement timing chain(s), gears, actuators, guides, and tensioners. Check that the timing marks align.

☐ **8.** Reassemble the front of the engine and torque all fasteners to factory specifications.

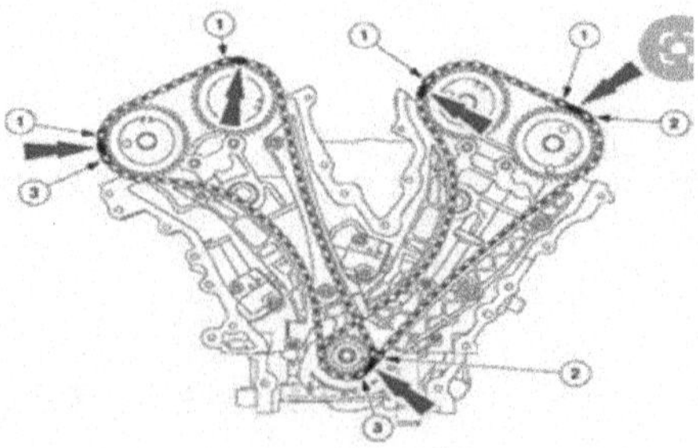

Pearson

Education Foundation

Valve Adjustment

Meets ASE Task: A1 – B-5 – P-2

Evaluation (Enter number from 4, 3, 2, 1): ☐

Name: ☐ **Date:** ☐ **Time on Task:** ☐

Make/Model/Year: ☐ **VIN:** ☐

☐ **1.** Check service information and determine the specified procedure to follow to adjust the valves. List the procedure: ☐

☐ **2.** The engine is equipped with ☐ ☐ Mechanical lifters ☐ ☐ Hydraulic Lifters.

☐ **3.** Compare the original valve clearance or adjustment to the specified clearance.

		Original	Specified
Cylinder #1	**Intake**		
	Exhaust		
Cylinder #2	**Intake**		
	Exhaust		
Cylinder #3	**Intake**		
	Exhaust		
Cylinder #4	**Intake**		
	Exhaust		
Cylinder #5	**Intake**		
	Exhaust		
Cylinder #6	**Intake**		
	Exhaust		

Cylinder #7	**Intake**		
	Exhaust		
Cylinder #8	**Intake**		
	Exhaust		

Pearson

Education Foundation

Module Identification and Status

Meets ASE Task: A-8 – B-1 – P-1

Evaluation (Enter number from 4, 3, 2, 1): ☐

Name: ☐ **Date:** ☐ **Time on Task:** ☐

Make/Model/Year: ☐ **VIN:** ☐

☐ **1.** Check service information for the procedure and scan tool configuration needed to identify the modules that are on the vehicle network.

☐ **2.** Using the specified scan tool and procedures, determine the modules that are on the vehicle being tested.

Module Name	Purpose of Module

☐ **3.** What is the status of the modules? ☐

☐ **4.** Are there any modules missing from the scan tool display that are known to be on the vehicle?

☐ Yes ☐ No (describe) ____________________

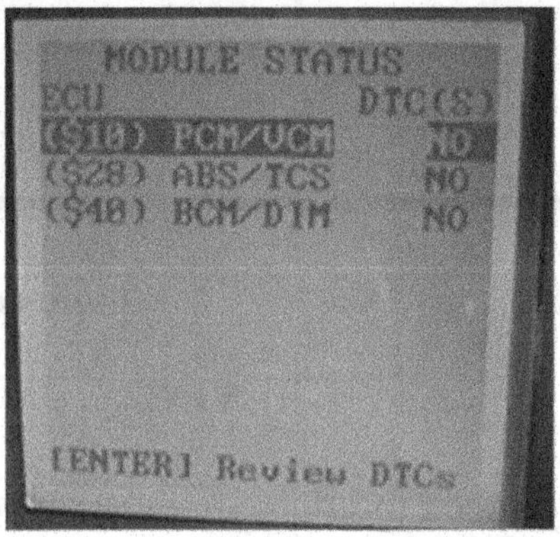

Pearson

Education Foundation

Module Communication

Meets ASE Task: A6 - G-8 – P-1

Evaluation (Enter number from 4, 3, 2, 1): ☐

Name: ☐ **Date:** ☐ **Time on Task:** ☐

Make/Model/Year: ☐ **VIN:** ☐

☐ **1.** Check service information for the specified method and procedures to follow to check for proper module communications.

☐ **2.** Check all that apply:

- ☐ ☐ Use a factory scan tool.
- ☐ ☐ Use a generic OBD-II scan tool.
- ☐ ☐ Use a DMM to check for resistance on communication circuits.
- ☐ ☐ Use a fused jumper wire and a scan tool to diagnose communication errors.
- ☐ ☐ Other (describe) ☐

☐ **3.** List the modules that are present in the vehicle being tested.

a.	f.
b.	g.
c.	h.
d.	j.
e.	k.

☐ **4.** Based on the test results, what is the necessary action?

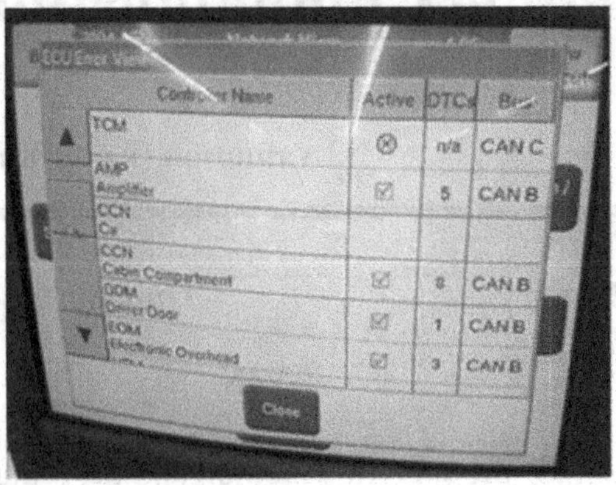

Pearson

Education Foundation

OBD II Connector Identification

Meets ASE Task: A8 – A-2 – P-1

Evaluation (Enter number from 4, 3, 2, 1): ☐

Name: ☐ **Date:** ☐ **Time on Task:** ☐

Make/Model/Year: ☐ **VIN:** ☐

☐ **1.** Check service information and check which cavities of the OBD II diagnostic link connector (DLC) have electrical (metal) terminals.

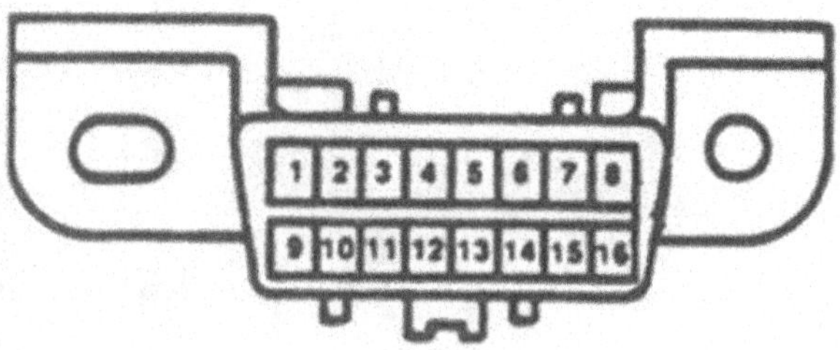

☐ **2.** Use service information and determine the identification for each of the terminals.

1. ☐
2. ☐
3. ☐
4. ☐
5. ☐
6. ☐
7. ☐
8. ☐
9. ☐
10. ☐
11. ☐

12.

13.

14.

15.

16.

Pearson

Education Foundation

Retrieving OBD II Diagnostic Trouble Codes

Meets ASE Task: A8 – A-2 – P-1

Evaluation (Enter number from 4, 3, 2, 1): ☐

Name: ☐ **Date:** ☐ **Time on Task:** ☐

Make/Model/Year: ☐ **VIN:** ☐

☐ **1.** A scan tool is required to retrieve diagnostic trouble codes from an OBD II vehicle.
Every OBD II scan tool will be able to read all generic Society of Automotive Engineers (SAE) DTCs from any vehicle.

☐ **2.** Retrieve the DTCs using a scan tool.

Specify which scan tool was used = ☐

DTCs ☐ ☐ ☐ ☐ ☐

☐ **3.** If no DTCs are displayed, set a DTC by disconnecting a sensor such as the throttle position (TP) sensor and then starting and running the engine.

☐ **4.** Did the scan tool display both a generic OBD II (Poxxx) code and a manufacturer's specific DTC (P1xxx) code?

Yes ☐ ☐ No ☐ ☐

☐ **5.** Clear the stored DTCs using the scan tool.

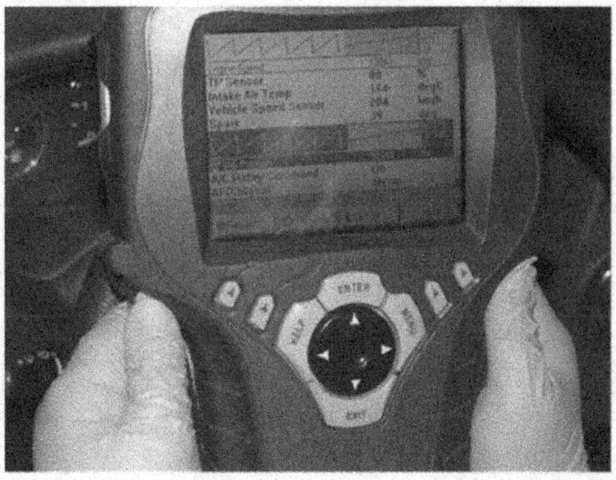

Pearson

Education Foundation

OBD II Monitors Status

Meets ASE Task: A8 – A-2 – P-1

Evaluation (Enter number from 4, 3, 2, 1): ☐

Name: ☐ **Date:** ☐ **Time on Task:** ☐

Make/Model/Year: ☐ **VIN:** ☐

☐ **1.** Check service information for the specified procedures and test equipment needed to determine the status of the inspection and maintenance (I/M) monitors.

☐ **2.** Connect a scan tool and check the status of the I/M monitors and list their status.

Monitor	Status

☐ **3.** Check service information and determine under what condition the vehicle must be driven to have the monitors run.

☐ **4.** Based on the results, what is the needed action?

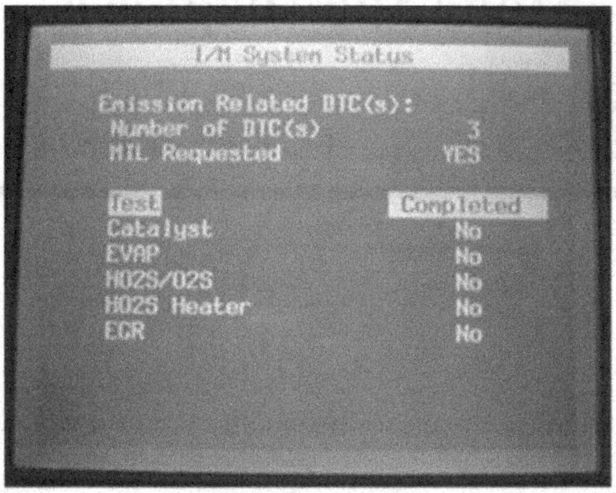

Pearson

Education Foundation

Mode $06 Diagnosis

Meets ASE Task: A6 – B-4 – P-1

Evaluation (Enter number from 4, 3, 2, 1): ☐

Name: ☐ **Date:** ☐ **Time on Task:** ☐

Make/Model/Year: ☐ **VIN:** ☐

☐ **1.** Check service information for the prescribed method to follow to verify that all monitors have run and passed. Describe the specified procedure.

☐ **2.** What scan tool was used to access mode $06 data?

☐ **3.** Which monitors have not run?

☐ **4.** What monitors have run?

☐ **5.** Which, if any, monitors failed?

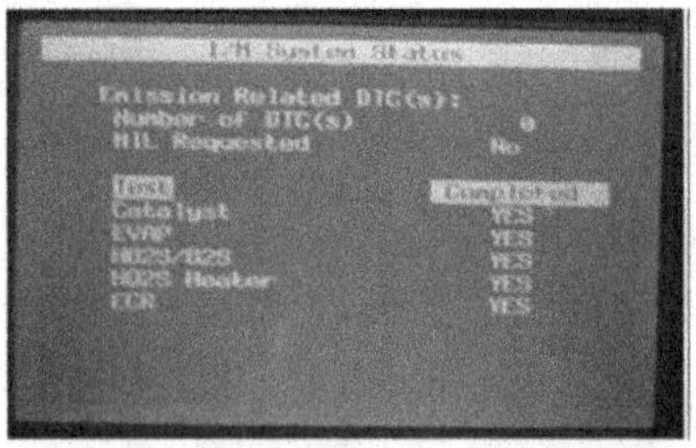

Pearson

Education Foundation

Battery, Starting, and Charging Systems

Meets ASE Task: A8 – A-12 – P-1

Evaluation (Enter number from 4, 3, 2, 1): ☐

Name: ☐ **Date:** ☐ **Time on Task:** ☐

Make/Model/Year: ☐ **VIN:** ☐

☐ **1.** Perform a battery state-of-charge test.

- ☐ What tester was used? ☐
- ☐ Describe the results. ☐
- ☐ Based on the test results, what is the needed action? ☐

☐ **2.** Perform a starter draw test.

- ☐ Describe the results. ☐
- ☐ Based on the test results, what is the needed action? ☐

☐ **3.** Perform a charging system output test.

- ☐ Describe the results. ☐
- ☐ Based on the test results, what is the needed action? ☐

Pearson

Education Foundation

Immobilizer System Diagnosis

Meets ASE Task: A-8 – B-9 – P-2

Evaluation (Enter number from 4, 3, 2, 1): ☐

Name: ☐ **Date:** ☐ **Time on Task:** ☐

Make/Model/Year: ☐ **VIN:** ☐

☐ **1.** An immobilizer system can cause the following conditions: (check all that apply)

- ☐ No crank condition (starter motor does not operate).
- ☐ Engine cranks but does not start.
- ☐ The engine starts but stops almost immediately.
- ☐ Other (describe): ☐

☐ **2.** Does the customer complaint match any of the above? ☐ Yes ☐ No

☐ **3.** Is the security light on? ☐ Yes ☐ No

☐ **4.** A valid key must be used. Is there another key? ☐ Yes ☐ No

If yes, try using the other key. ☐ Normal operation ☐ Still does not start.

☐ **5.** Any stored diagnostic trouble codes? List stored codes:

☐ **6.** Check service information for the specified procedure to follow when diagnosing an immobilizer fault. Describe the specified procedure:

☐ **7.** What is the necessary action?

Pearson

Education Foundation

Ignition System Identification

Meets ASE Task: A8 – A-1 – P-1

Evaluation (Enter number from 4, 3, 2, 1): ☐

Name: ☐ **Date:** ☐ **Time on Task:** ☐

Make/Model/Year: ☐ **VIN:** ☐

☐ **1.** Check service information and determine what type of ignition system is used on this vehicle?

- ☐ Distributor ignition (DI)
- ☐ Waste-spark (EI)
- ☐ Coil-on-plug
- ☐ Other (describe): ☐

☐ **2.** Check service information and determine what type of primary circuit switching device is used on this system.

- ☐ Pickup coil (pulse generator)
- ☐ Hall-effect sensor
- ☐ Magnetic sensor
- ☐ Optical sensor
- ☐ Other (describe): ☐

☐ **3.** What color wires are used on the switching device?

☐ **4.** Using service information, determine where the primary ignition switching device signal goes:

☐ Ignition control module (ICM)

☐ Computer (PCM)

☐ Other (describe): ____________________

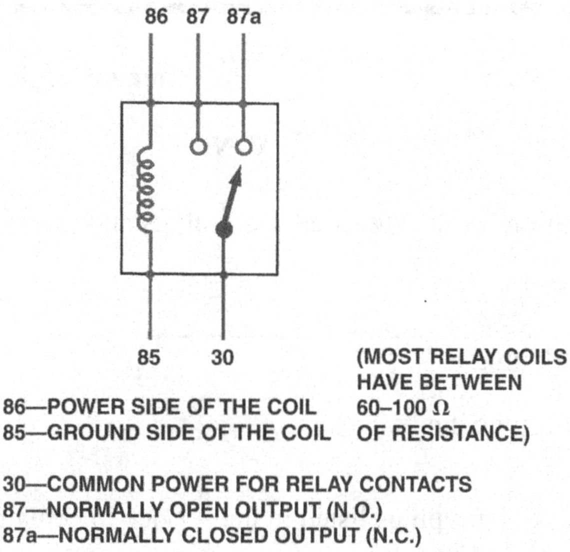

Pearson

Education Foundation

Spark Plug Specifications

Meets ASE Task: A8 – A-1 – P-1

Evaluation (Enter number from 4, 3, 2, 1): ☐

Name: ☐ **Date:** ☐ **Time on Task:** ☐

Make/Model/Year: ☐ **VIN:** ☐

☐ **1.** Check service information for the specified spark plugs and the recommended spark plug gap.

☐ Recommended spark plug = ☐

☐ Recommended spark plug gap = ☐

☐ **2.** Are there any alternative spark plugs listed in the service information? If so, list the numbers or identifying information.

a. ☐

b. ☐

c. ☐

d. ☐

e. ☐

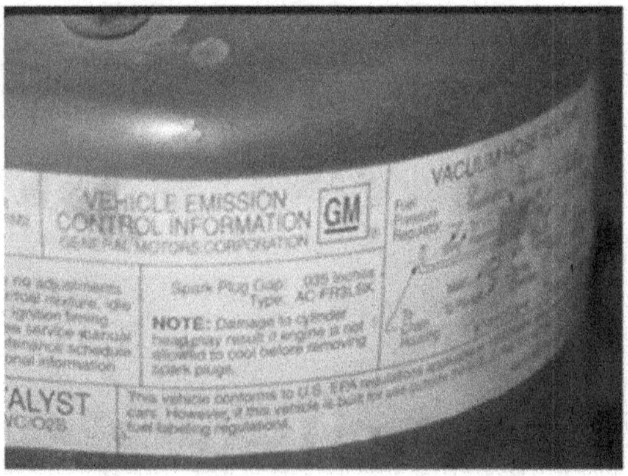

Pearson

Education Foundation

Electronic Ignition Diagnosis

Meets ASE Task: A8 – C-3 – P-1

Evaluation (Enter number from 4, 3, 2, 1): ☐

Name: ☐ **Date:** ☐ **Time on Task:** ☐

Make/Model/Year: ☐ **VIN:** ☐

☐ **1.** Check service information for the specified diagnostic procedures to follow when troubleshooting the ignition system.

☐ **2.** Most test procedures specify that the spark be tested using a spark tester. Use a spark tester and determine that a spark does occur at each cylinder.

Cylinder #1	☐	OK ☐	☐	NOT OK ☐	☐	(describe)	☐
Cylinder #2	☐	OK ☐	☐	NOT OK ☐	☐	(describe)	☐
Cylinder #3	☐	OK ☐	☐	NOT OK ☐	☐	(describe)	☐
Cylinder #4	☐	OK ☐	☐	NOT OK ☐	☐	(describe)	☐
Cylinder #5	☐	OK ☐	☐	NOT OK ☐	☐	(describe)	☐
Cylinder #6	☐	OK ☐	☐	NOT OK ☐	☐	(describe)	☐
Cylinder #7	☐	OK ☐	☐	NOT OK ☐	☐	(describe)	☐
Cylinder #8	☐	OK ☐	☐	NOT OK ☐	☐	(describe)	☐

☐ **3.** Based on the test results, what is the needed action?

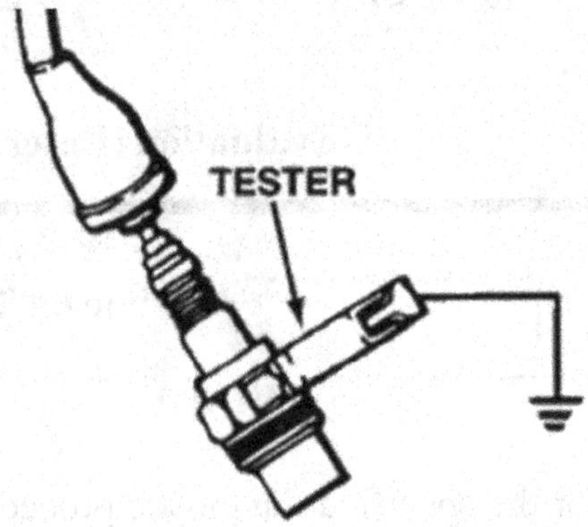

Pearson

Education Foundation

Scan Tool Testing of the Ignition System

Meets ASE Task: A8 – B-5 – P-1

Evaluation (Enter number from 4, 3, 2, 1): ☐

Name: ☐ **Date:** ☐ **Time on Task:** ☐

Make/Model/Year: ☐ **VIN:** ☐

☐ **1.** Check service information for the recommended procedures to follow when using a scan tool to monitor what the vehicle computer is looking at and what commands are being sent to the various engine actuators and components.

☐ **2.** Connect the scan tool to the data link connector (DLC) of the vehicle and scroll through the various parameters until engine RPM and ignition spark timing can be viewed.

Idle RPM = ☐ Spark advance at idle = ☐

☐ **3.** Slowly increase engine speed and observe the amount of spark advance.

Spark advance at 1000 RPM = ☐ degrees

Spark advance at 1500 RPM = ☐ degrees

Spark advance at 2000 RPM = ☐ degrees

Spark advance at 2500 RPM = ☐ degrees

☐ **4.** Scroll the display of the scan tool until knock sensor (KS) activity or timing retard is displayed (if the vehicle is so equipped).

KS signal at idle = ☐ (should be near zero)

☐ **5.** Increase engine speed while observing KS or timing retard amount. Did the computer retard timing?

Yes ☐ ☐ No ☐ ☐

☐ **6.** Lightly tap on the engine block and observe KS or timing retard with the engine warmed above idle speed. Was a knock detected?

Yes ☐ No ☐

☐ **7.** Based on the test results, what is the needed action?

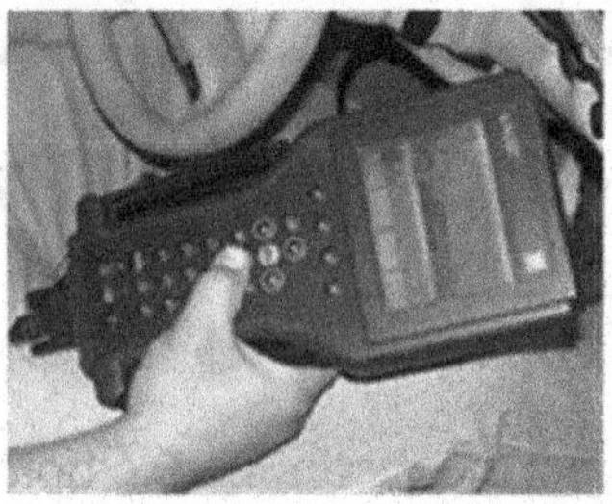

Pearson

Education Foundation

Ignition Scope Analysis

Meets ASE Task: A8 – C-3 – P-1

Evaluation (Enter number from 4, 3, 2, 1): ☐

Name: ☐ **Date:** ☐ **Time on Task:** ☐

Make/Model/Year: ☐ **VIN:** ☐

☐ **1.** Check service information regarding the specified method for attaching and using a secondary circuit oscilloscope.

☐ **2.** Type of ignition:

☐ ☐ Distributor ☐ ☐ Waste spark ☐ ☐ Coil-on-plug

☐ **3.** Connect the ignition scope to the system as per the scope manufacturer's instructions.

☐ **4.** Start the engine and observe the secondary ignition waveform.

	Firing Voltage (KV)	Spark Line Length (ms)
Cylinder #1		
Cylinder #2		
Cylinder #3		
Cylinder #4		
Cylinder #5		
Cylinder #6		

Cylinder #7 | |

Cylinder #8 | |

☐ **5.** Based on the test results, what is the needed action?

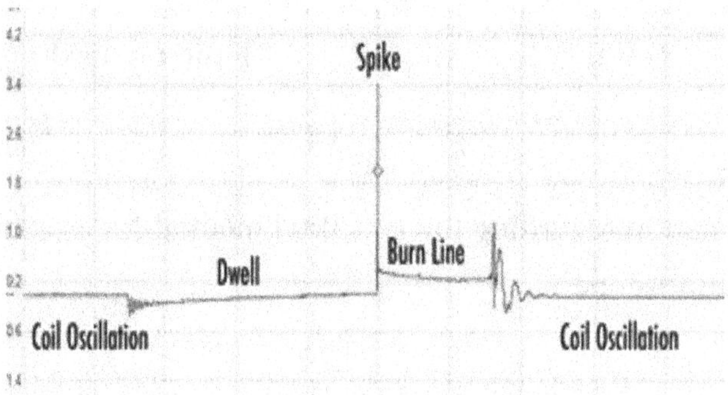

Pearson

ASE Education Foundation

Spark Plugs Inspection

Meets ASE Task: A8 – C-2 – P-1

Evaluation (Enter number from 4, 3, 2, 1): ☐

Name: ☐ **Date:** ☐ **Time on Task:** ☐

Make/Model/Year: ☐ **VIN:** ☐

☐ **1.** Check service information and determine the correct plug code and gap for your vehicle using a spark plug application guide.

Brand ☐ Code # ☐ Gap ☐

☐ **2.** Remove and label all the coils or spark plug wires.

☐ **3.** Determine the condition and gap of all spark plugs:

	Condition	Gap
1.		
2.		
3.		
4.		
5.		
6.		
7.		
8.		

☐ **4.** Reinstall the spark plug (start by hand).

☐ **5.** Use a torque wrench and torque the spark plugs to the proper torque.

Specified torque = ☐

☐ **6.** Reinstall the coils or spark plug wires. Start engine and check for roughness.

☐ **7.** Based on the inspection of the spark plugs, what is the needed action?

Pearson

Education Foundation

Ignition Coil Testing

Meets ASE Task: A8 – C-3 – P-1

Evaluation (Enter number from 4, 3, 2, 1): ☐

Name: ☐ **Date:** ☐ **Time on Task:** ☐

Make/Model/Year: ☐ **VIN:** ☐

☐ **1.** Check service information for the specified ignition coil testing procedure.

☐ **2.** Visually inspect the coil(s) for carbon track and other faults. OK ☐ NOT OK ☐

☐ **3.** Check the primary winding resistance.

Specification: ☐

Actual: ☐

OK ☐ NOT OK ☐

☐ **4.** Check the secondary winding resistance.

Specification: ☐

Actual: ☐

OK ☐ NOT OK ☐

☐ **5.** Perform resistance checks for short-to-ground.

OK ☐ NOT OK ☐

☐ **6.** Based on the inspection and tests, what is the needed action?

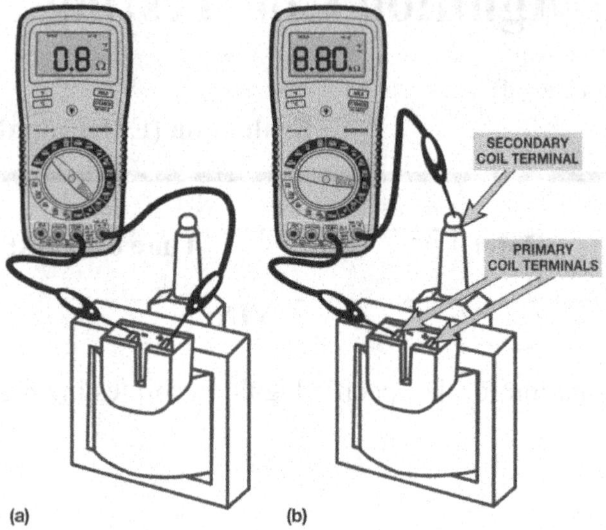

Pearson

ASE Education Foundation

Position Sensor Diagnosis

Meets ASE Task: A8 – C-4 – P-1

Evaluation (Enter number from 4, 3, 2, 1): ☐

Name: ☐ **Date:** ☐ **Time on Task:** ☐

Make/Model/Year: ☐ **VIN:** ☐

☐ **1.** Check service information for the type of sensor used for the crankshaft position (CKP) and the camshaft position (CMP).

☐ CKP – Describe the type of sensor ☐

☐ CMP – Describe the type of sensor ☐

☐ **2.** Check service information and determine the recommended testing procedures for testing the CKP sensor.

Wire colors and their purpose:

Wire #1 ☐ ☐

Wire #2 ☐ ☐

Wire #3 ☐ ☐

Recommended testing procedure and specifications: ☐

Test results: ☐ OK ☐ ☐ NOT OK ☐ ☐

☐ **3.** Check service information and determine the recommended testing procedures for testing the CMP sensor.

Wire colors and their purpose:

Wire #1 ☐ ☐

Wire #2 ☐ ☐

Wire #3

Recommended testing procedure and specifications:

Test results: OK ☐ NOT OK ☐

☐ **4.** Based on the test results, what is the needed action?

Pearson

Education Foundation

Position Sensor Waveform Testing

Meets ASE Task: A8 – C-4 – P-1

Evaluation (Enter number from 4, 3, 2, 1): ☐

Name: ☐ **Date:** ☐ **Time on Task:** ☐

Make/Model/Year: ☐ **VIN:** ☐

☐ **1.** Check service information regarding the location and type of position sensors used on the vehicle being tested.

	Crankshaft Position (CKP)	Camshaft Position (CMP)
Location (describe):		
Type of sensor:		

☐ **2.** Check service information for the wire colors used and their purpose.

	Crankshaft Position (CKP)	Camshaft Position (CMP)
Wire 1 (color and purpose):		
Wire 2 (color and purpose):		
Wire 3 (color and purpose):		

☐ **3.** Check service information for the specified voltage output and/or waveform for the CKP and CMP sensors. CKP ☐ CMP ☐

☐ **4.** Following the test equipment manufacturer's instructions, capture the waveforms.

CKP – OK ☐ ☐ CMP – OK ☐ ☐

☐ **5.** Based on the comparison between the captured waveform and the specified waveform, what action is needed?

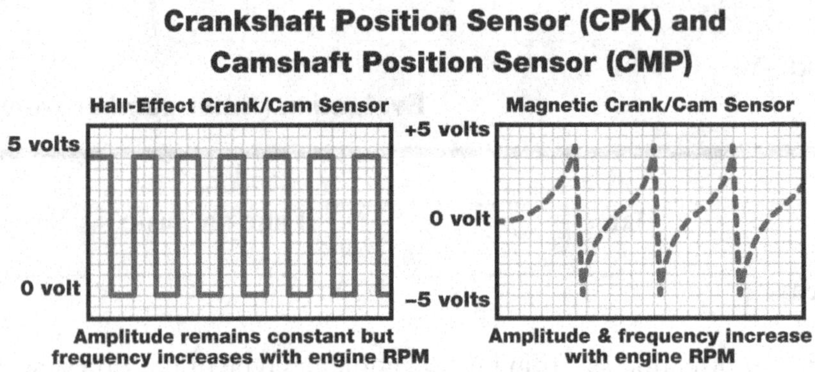

Pearson

Education Foundation

Engine Problem Analysis

Meets ASE Task: A8 – A-7 – P-2

Evaluation (Enter number from 4, 3, 2, 1): ☐

Name: ☐ **Date:** ☐ **Time on Task:** ☐

Make/Model/Year: ☐ **VIN:** ☐

☐ **1.** State the customer's concern regarding the engine problem. ☐

- ☐ Excessive noise? ☐
- ☐ Exhaust smoke? ☐
- ☐ Engine operation (missing, runs rough, etc.)? ☐
- ☐ Other (describe): ☐

☐ **2.** Based on the symptoms described above, what tests should be performed?

- ☐ ☐ Compression test
- ☐ ☐ Cylinder linkage test
- ☐ ☐ Power balance test
- ☐ ☐ Other (describe): ☐

☐ **3.** Based on the symptoms and the test results, what is the needed action?

Pearson

Education Foundation

Fluid Leakage Detection

Meets ASE Task: A1 – A-4 – P-1

Evaluation (Enter number from 4, 3, 2, 1): ☐

Name: ☐ **Date:** ☐ **Time on Task:** ☐

Make/Model/Year: ☐ **VIN:** ☐

☐ **1.** Review the service information and describe the process for checking the engine for fuel, oil, and coolant leaks.

- ☐ List any special tools required ☐
- ☐ List any bulletins that may apply ☐

☐ **2.** Inspect the engine for fuel, oil, and coolant leaks. Describe the results.

☐ **3.** Based on the inspection described above, what action is needed

Pearson

Education Foundation

Engine Noise and Vibration Diagnosis

Meets ASE Task: A8 – A-6 – P-2

Evaluation (Enter number from 4, 3, 2, 1): ☐

Name: ☐ **Date:** ☐ **Time on Task:** ☐

Make/Model/Year: ☐ **VIN:** ☐

☐ **1.** When is the engine noise or vibration detected? Check all that apply.

- ☐ At engine start (warm engine)
- ☐ At engine start (cold engine)
- ☐ Idle (cold)
- ☐ Idle (hot)
- ☐ During acceleration
- ☐ Under heavy load
- ☐ Under light load
- ☐ During deceleration
- ☐ Other (describe): ☐

☐ **2.** Describe the noise or vibration. Check all that apply.

- ☐ Clicking sound (like a ball-point pen)
- ☐ Clacking noise (like tapping on metal)
- ☐ Rattle (like a baby rattle)
- ☐ Clatter (like rolling marbles)
- ☐ Whine (like an electric motor running)
- ☐ Clunk (like a door closing)
- ☐ Other (describe): ☐

☐ **3.** Check service information regarding the noise and the associated tests and procedures that should be followed. What action is needed?

Pearson

Education Foundation

Vacuum Testing

Meets ASE Task: A8 – A-8 – P-1

Evaluation (Enter number from 4, 3, 2, 1): ☐

Name: ☐ **Date:** ☐ **Time on Task:** ☐

Make/Model/Year: ☐ **VIN:** ☐

☐ **1.** Connect the vacuum gauge to a manifold vacuum source (source of vacuum at idle).

☐ **2.** Vacuum at idle = ☐ in. Hg. (Should be 17-21 in. Hg. and steady).

☐ **3.** Drive the vehicle on a level road in high gear at a steady speed.

Cruise vacuum = ☐ in. Hg. (Should be 10 - 15 in. Hg.)

☐ **4.** Accelerate the vehicle in high gear to W.O.T.

W.O.T. vacuum = ☐ in. Hg. (Should be almost zero)

☐ **5.** Decelerate the vehicle from 50 MPH with the throttle closed.

Deceleration vacuum = ☐ in. Hg. (Should be higher than idle vacuum)

☐ **6.** With the engine operating and the transmission in neutral or park, and the brake firmly applied, raise the engine speed to 2,000 RPM and hold for one full minute. This tests for an exhaust restriction. Results = ☐ in. Hg.

☐ **7.** Stop the engine. Disable the ignition. Crank the engine and observe the vacuum during cranking. Cranking vacuum = ☐ in. Hg. (Should be higher than 2.5 in. Hg.)

OK ☐ ☐ NOT OK ☐ ☐

☐ **8.** Based on the vacuum test results, what is the needed action?

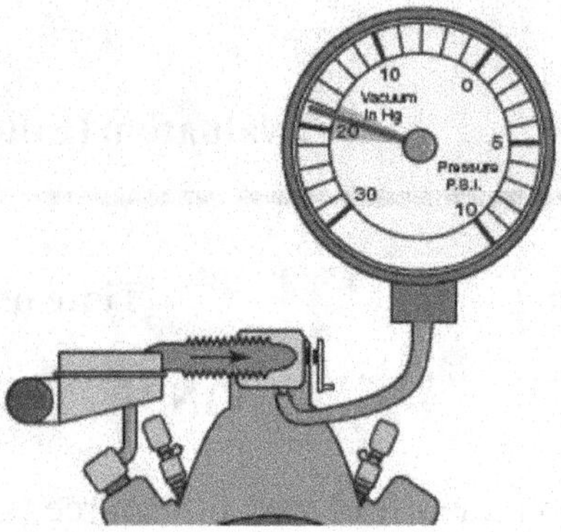

Pearson

Education Foundation

Cylinder Power Balance Test

Meets ASE Task: A8 – A-9 – P-1

Evaluation (Enter number from 4, 3, 2, 1): ☐

Name: ☐ **Date:** ☐ **Time on Task:** ☐

Make/Model/Year: ☐ **VIN:** ☐

☐ **1.** An automotive diagnostic scan tool or digital storage oscilloscope with relative compression can be used to determine cylinder balance. (Check the tool used)

☐ Scan Tool

☐ Digital storage oscilloscope

☐ Other (describe):

☐ **2.** Follow the equipment manufacturers' instructions and connect the tester to the engine. Instructions to connect to the engine include:

☐ **3.** Start the engine and allow it to reach normal operating temperature.

☐ **4.** Follow the instructions of the test equipment manufacturer and perform a cylinder power balance test. Record the results.

Cylinder #1 =	Cylinder #5 =
Cylinder #2 =	Cylinder #6 =
Cylinder #3 =	Cylinder #7 =
Cylinder #4 =	Cylinder #8 =

☐ **5.** If performing an engine speed (RPM) drop test, all cylinders should be within 50 RPM.

☐ **OK** ☐ **NOT OK (describe results)** ☐ **NA**

☐ **6.** If relative compression is being performed, all cylinders should be within 10%.

☐ **OK** ☐ **NOT OK (describe results)** ☐ **NA**

Pearson

Education Foundation

Compression Testing

Meets ASE Task: A8 – A-10 – P-1

Evaluation (Enter number from 4, 3, 2, 1): ☐

Name: ______ **Date:** ______ **Time on Task:** ______

Make/Model/Year: ______ **VIN:** ______

☐ **1.** Remove all spark plugs and disable the ignition system to avoid possible ignition coil damage.

☐ **2.** Block open the throttle (Refer to manufacturer's procedures if equipped with ETC).

☐ **3.** Crank the engine at least 4 "puffs" (compression strokes) while observing the gauge.

NOTE: For accurate test results, the engine should be at normal operating temperature.

RESULTS: **1st puff / final reading** **1st puff / final reading**

	1st puff		final reading		1st puff		final reading
1.		/		5.		/	
2.		/		6.		/	
3.		/		7.		/	
4.		/		8.		/	

NOTE: The 1st "puff" should be more than one-half of the pressure of the final puff. If the 1st puff is low, worn piston rings are likely – repeat the test.

☐ **4.** Reinstall all spark plugs except one. Perform a running compression test at idle and at 2000 RPM for each cylinder.

	Idle		2000 RPM
1.		1.	
2.		2.	
3.		3.	
4.		4.	
5.		5.	

6. [ ] 6. [ ]

7. [ ] 7. [ ]

8. [ ] 8. [ ]

☐ **5.** Based on the test results, what is the necessary action? [ ]

Pearson

Education Foundation

Cylinder Leakage Testing

Meets ASE Task: A8 – A-11 – P-1

Evaluation (Enter number from 4, 3, 2, 1): ☐

Name: ☐ **Date:** ☐ **Time on Task:** ☐

Make/Model/Year: ☐ **VIN:** ☐

☐ **1.** The engine should be at normal operating temperature.

☐ **2.** Rotate the engine until the piston of the cylinder being tested is at TDC on the compression stroke.

☐ **3.** Calibrate the cylinder leakage gauge.

☐ **4.** Install compressed air in the cylinder. Read the gauge. ☐ % of leakage

Check one:

☐ ☐ Good - less than 10%

☐ ☐ Acceptable - less than 20%

☐ ☐ Unacceptable - higher than 20%

☐ **5.** Check the location of air leakage:

☐ ☐ a. **radiator** - possible blown head gasket or cracked cylinder head.

☐ ☐ b. **tail pipe** - defective exhaust valve(s).

☐ ☐ c. **Throttle Body or air inlet** - defective intake valve(s).

☐ ☐ d. **oil filler cap** - possible worn or defective piston rings.

☐ **6.** Based on the test results, what is the necessary action?

Pearson

Education Foundation

Oil Pressure Measurement

Meets ASE Task: A1 – D-11 – P-1

Evaluation (Enter number from 4, 3, 2, 1): ☐

Name: ☐ **Date:** ☐ **Time on Task:** ☐

Make/Model/Year: ☐ **VIN:** ☐

☐ **1.** Locate and remove the sending unit using the proper size sending unit socket or wrench.

☐ **2.** Thread a mechanical oil pressure gauge into the hole where the sending unit was located.

☐ **3.** Route the oil pressure gauge hose away from the moving components of the engine.

☐ **4.** Start the engine and check for leaks.

☐ **5.** Record the oil pressure:

- ☐ Oil pressure @ idle ☐
- ☐ Oil pressure @ 1,000 RPM ☐
- ☐ Oil pressure @ 2,000 RPM ☐
- ☐ Oil pressure @ 3,000 RPM ☐

☐ **6.** Results: (check one)

- ☐ Great ☐ (over 10 psi per 1,000 RPM)
- ☐ Good ☐ (at 10 psi per 1,000 RPM)
- ☐ Bad ☐ (less than 10 psi per 1,000 RPM)

☐ **7.** Based on the test results, what is the needed action?

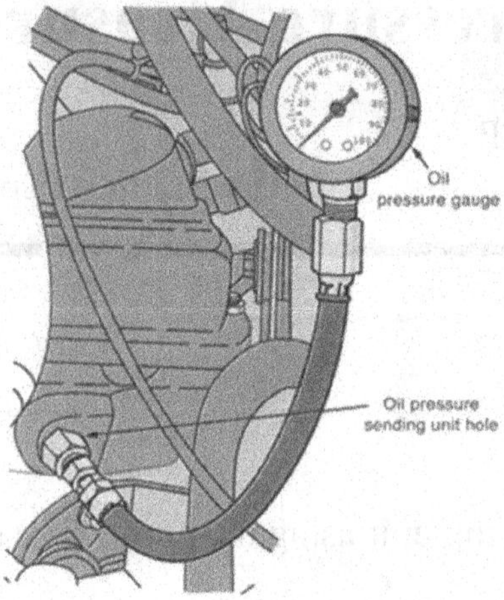

Pearson

Education Foundation

VVT Inspect and Replace Timing Chain(s) and Gears

Meets ASE Task: A1 – B-6 – P-1

Evaluation (Enter number from 4, 3, 2, 1):

Name: **Date:** **Time on Task:**

Make/Model/Year: **VIN:**

☐ **1.** Check service information to determine the specified procedure and special tools needed to inspect and replace the timing chain(s), gears, actuators, guides, and tensioners. List the special tools needed:

☐ **2.** Remove the engine front cover following the manufacturer's procedures.

☐ **3.** Using the original crankshaft bolt, rotate the engine until the crankshaft gear is in the desired location.

☐ **4.** Verify the timing marks on the variable cam actuators are in the correct location.

☐ **5.** Inspect the chain(s), gears, actuators, guides, and tensioners for wear.

☐ **6.** Based on the inspection, what components need to be replaced?

☐ **7.** Install the replacement timing chain(s), gears, actuators, guides, and tensioners. Check that the timing marks align.

☐ **8.** Reassemble the front of the engine and torque all fasteners to factory specifications.

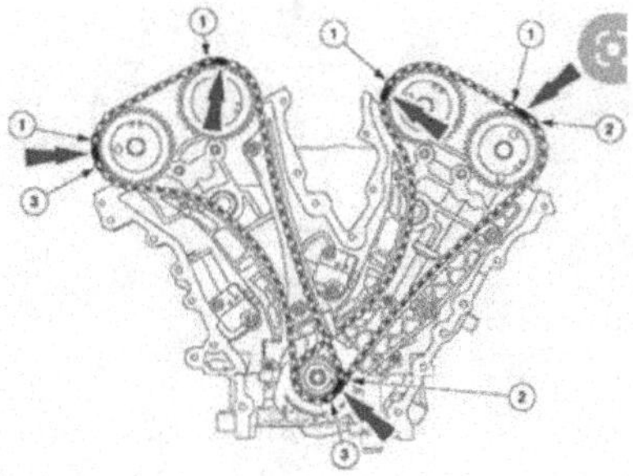

Pearson

Education Foundation

Variable Valve Timing System Diagnosis

Meets ASE Task: A8 – A-4 – P-1

Evaluation (Enter number from 4, 3, 2, 1): ☐

Name: ☐ **Date:** ☐ **Time on Task:** ☐

Make/Model/Year: ☐ **VIN:** ☐

☐ **1.** What diagnostic trouble code (DTC) has been set? (Check all that apply)

- ☐ ☐ P0010 – "A" camshaft position actuator circuit
- ☐ ☐ P0011 – "A" camshaft position timing or system performance
- ☐ ☐ P0012 – "A" camshaft position – timing over-retarded
- ☐ ☐ P0013 – "B" camshaft position – actuator circuit
- ☐ ☐ P0014 – "B" camshaft position – timing or system performance
- ☐ ☐ P0015 – "B" camshaft position – timing over-retarded
- ☐ ☐ P0016 – P0019 – crankshaft position – camshaft position correlation error
- ☐ ☐ Other (describe) ☐

☐ **2.** What is the specified diagnostic step to follow according to service information?

☐ **3.** What was the root cause of the problem?

Pearson

Education Foundation

Temperature Sensor Scan Tool Diagnosis

Meets ASE Task: A8 – B-5 – P-1

Evaluation (Enter number from 4, 3, 2, 1): ☐

Name: ☐ **Date:** ☐ **Time on Task:** ☐

Make/Model/Year: ☐ **VIN:** ☐

☐ **1.** Check service information for the recommended method for checking temperature sensors using a scan tool.

☐ **2.** List all the temperature sensors that are on the vehicle. Check all that apply.

- ☐ ☐ ECT (describe location) ☐
- ☐ ☐ IAT (describe location) ☐
- ☐ ☐ TFT (describe location) ☐
- ☐ ☐ Other (describe) ☐

☐ **3.** On a vehicle that has not been operated for several hours, use a scan tool, and list the temperature displayed for various temperature sensors.

ECT = ☐

IAT = ☐

TFT = ☐

Other (describe) = ☐

☐ **4.** Based on the test results, what is the needed action?

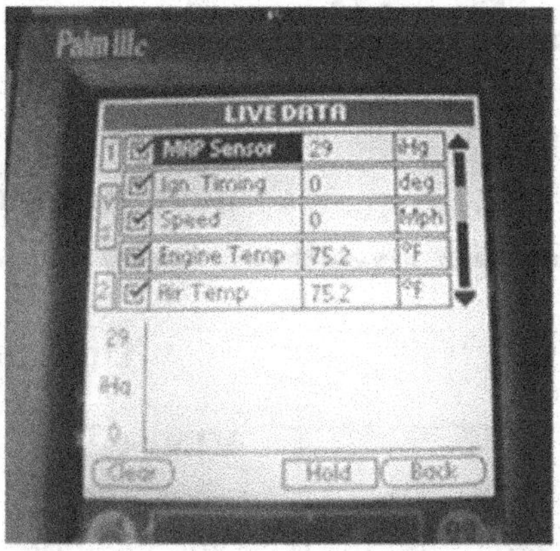

Pearson

Education Foundation

Throttle Position Sensor Testing

Meets ASE Task: A8 – B-5 – P-1

Evaluation (Enter number from 4, 3, 2, 1):

Name: **Date:** **Time on Task:**

Make/Model/Year: **VIN:**

☐ **1.** Check service information regarding the type and location of the throttle position sensor used on the vehicle being tested.

☐ **2.** Check service information for the wire colors used and their purpose.

Wire 1 (color and purpose):

Wire 2 (color and purpose):

Wire 3 (color and purpose):

☐ **3.** Measure the reference voltage =

☐ **4.** Measure the sensor signal voltage at closed throttle =

☐ **5.** Measure the TP sensor at wide-open throttle (W.O.T.) =

☐ **6.** Measure the TP signal sweep using an oscilloscope. Describe the results.

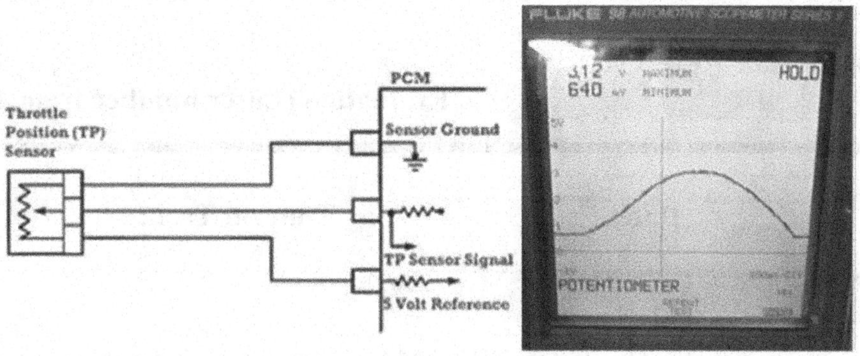

Pearson

Education Foundation

MAP Sensor Diagnosis

Meets ASE Task: A8 – A-5 – P-1

Evaluation (Enter number from 4, 3, 2, 1): ☐

Name: ☐ **Date:** ☐ **Time on Task:** ☐

Make/Model/Year: ☐ **VIN:** ☐

☐ **1.** Check service information for the specified MAP sensor diagnostic procedure.

☐ **2.** Perform a thorough visual inspection including:

a. Check the condition of vacuum hose (if equipped).
b. Check that the sensor is properly mounted/sealed at the intake manifold.
c. Check the condition of the electrical connector at the sensor.

☐ **3.** Turn the ignition key on (engine off), read and record the MAP sensor voltage (or frequency) = ☐ volts (Hz) (use either a scan tool or digital meter connected to the signal wire). (Should be about 4.60 to 4.80 volts or 156-159 Hz.)

OK ☐ ☐ NOT OK ☐ ☐

☐ **4.** Start the engine and operate until normal operating temperature is achieved. Read and record the MAP sensor voltage (or Hz) at idle speed = ☐ volts (Hz). (Should be between 0.9 and 1.6 volts (102-109 Hz) if the engine varies between 17 and 21 inches of Hg.)

OK ☐ ☐ NOT OK ☐ ☐

☐ **5.** Based on these tests, what is the needed action?

Pearson

Education Foundation

MAF Sensor Diagnosis

Meets ASE Task: A8 – B-5 – P-1

Evaluation (Enter number from 4, 3, 2, 1): ☐

Name: ☐ **Date:** ☐ **Time on Task:** ☐

Make/Model/Year: ☐ **VIN:** ☐

☐ **1.** Check service information for the specified procedure to follow to test the MAF sensor.

☐ **2.** Use a meter or scope with a frequency counter to record frequency or voltage at idle and at WOT (short bursts).

at idle = ☐ at WOT = ☐

☐ **3.** Use a scan tool and record grams per second.

at idle = ☐ at WOT = ☐

☐ **4.** A good MAF should read:

- greater than 100 grams per second (scan tool diagnosis)
- higher than 7000 Hertz (7 KHz) (digital MAF)
- higher than 4 volts (analog MAF)

☐ **5.** If the MAF sensor reading does not exceed these values, the sensing wire may be contaminated or the sensor itself is defective.

☐ **6.** Based on the test results, what is the needed action?

Pearson

Education Foundation

Electronic Throttle Control System ID

Meets ASE Task: A8 – A-1 – P-1

Evaluation (Enter number from 4, 3, 2, 1): ☐

Name: ☐ **Date:** ☐ **Time on Task:** ☐

Make/Model/Year: ☐ **VIN:** ☐

☐ **1.** Check service information to determine if the vehicle being serviced is equipped with an electronic throttle control (ETC) system. Describe how it was determined. (Check all that apply)

- ☐ Visual inspection
- ☐ Service information
- ☐ Scan tool
- ☐ Other (describe): ☐

☐ **2.** Does the electronic throttle control system use a cable between the accelerator pedal and the APP sensor?

☐ ☐ Yes ☐ ☐ No

☐ **3.** What is the relearn procedure that needs to be followed if the electronic throttle control system throttle body assembly is replaced? Describe the specified procedure.

Pearson

Education Foundation

Oxygen Sensor Diagnosis

Meets ASE Task: A8 – B-5 – P-1

Evaluation (Enter number from 4, 3, 2, 1): ☐

Name: ☐ **Date:** ☐ **Time on Task:** ☐

Make/Model/Year: ☐ **VIN:** ☐

☐ **1.** Connect the scan tool to the DLC and start the engine.

☐ **2.** Operate the engine at a fast idle (2500 RPM) for 2 minutes to allow time for the oxygen sensor to warm to operating temperature.

☐ **3.** Observe the oxygen sensor activity on the scan tool to verify closed loop operation.

☐ **4.** Select "snap shot" mode and hold the engine speed steady and start recording.

☐ **5.** Play back snap shot and place a mark beside each range of oxygen sensor voltage for each frame of the snap shot.

Between 0 and 300 mV	Between 300 and 600 mV	Between 600 and 1000 mV
(record # of times)	(record # of times)	(record # of times)

☐ **6.** Results: A good oxygen sensor and computer system should result in most snap shot values at both ends (0 to 300 and 600 to 1000 mV). If most of the readings are in the middle, the oxygen sensor is not working correctly.

OK ☐ ☐ NOT OK ☐

☐ **7.** Based on the test results, what is the needed action?

Pearson

Education Foundation

Wide-Band Oxygen Sensor

Meets ASE Task: A8 – B-5 – P-1

Evaluation (Enter number from 4, 3, 2, 1): ☐

Name: ☐ **Date:** ☐ **Time on Task:** ☐

Make/Model/Year: ☐ **VIN:** ☐

☐ **1.** Wide-band oxygen sensors use four, five, or six wires. Check service information for the color and identification of each of the wires used on the wide-band oxygen sensors.

	Wire Color	Purpose	Typical Voltage on Wire (CL)
a.			
b.			
c.			
d.			
e.			
f.			

☐ **2.** Check service information for the specified testing procedure for diagnosing the wide-band oxygen sensor.

☐ **3.** Check all the tools that that are specified in the test procedure:

- ☐ ☐ Scan tool
- ☐ ☐ DMM
- ☐ ☐ DSO
- ☐ ☐ Other (describe) ☐

☐ **4.** Based on the results of the specified test procedure, what is the needed action?

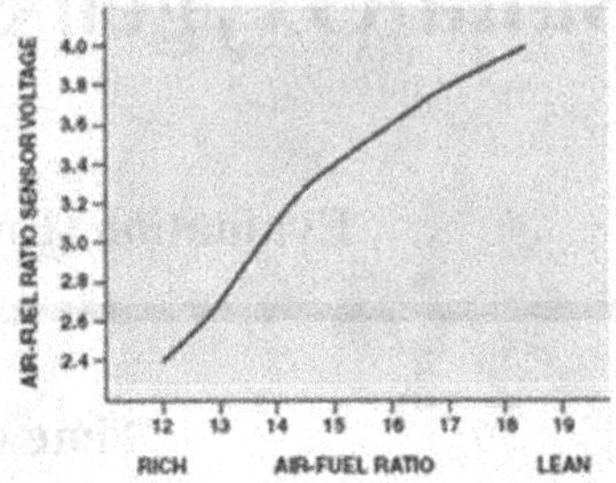

Pearson

Education Foundation

Fuel Trim Diagnosis

Meets ASE Task: A-8 – B-8 – P-1, A-8 – D-13 – P-2

Evaluation (Enter number from 4, 3, 2, 1):

Name: **Date:** **Time on Task:**

Make/Model/Year: **VIN:**

☐ **1.** Check service information for the recommended procedure to follow when diagnosing driveability or emission-related faults without any stored diagnostic trouble codes. Explain the recommended procedure.

☐ **2.** Using a scan tool, check the LTFT and STFT:

LTFT = STFT = OK ☐ NOT OK ☐

Explain possible causes:

☐ **3.** While driving the vehicle with a scan tool connected, record LTFT and STFT. Record any speed/load where LTFT or STFT exceeded 10%.

☐ **4.** Based on the results of the tests, what is the necessary action?

Pearson

Education Foundation

Fuel Pump Testing

Meets ASE Task: A8 – D-8 – P-1

Evaluation (Enter number from 4, 3, 2, 1): ☐

Name: ☐ **Date:** ☐ **Time on Task:** ☐

Make/Model/Year: ☐ **VIN:** ☐

☐ **1.** Check service information and determine the factory specifications for acceptable fuel pump pressure.

Fuel pump pressure specifications = ☐

☐ **2.** Check service information, locate the fuel system pressure test valve or port, and describe its location. If no test port, describe the process to adapt a pressure gauge to the system.

☐ **3.** Connect a fuel pressure gauge to the fuel pressure Schrader valve.

☐ **4.** Start the engine and observe the fuel pressure.

fuel pressure = ☐ OK ☐ ☐ NOT OK ☐ ☐

☐ **5.** If equipped; connect a hand-operated vacuum pump to the fuel pressure regulator and apply 20 in. Hg. of vacuum.

Did the pressure decrease? Yes ☐ ☐ No ☐ ☐

☐ **6.** Check fuel pump volume (0.5 to 1.0 gallons per minute).

☐ **7.** Based on the tests, what is the needed action?

Pearson

Education Foundation

Fuel Pump Current Draw Test

Meets ASE Task: A8 – D-8 – P-1

Evaluation (Enter number from 4, 3, 2, 1): ☐

Name: ☐ **Date:** ☐ **Time on Task:** ☐

Make/Model/Year: ☐ **VIN:** ☐

☐ **1.** Many electric fuel pumps can be measured for current draw in amperes. An abnormal amperage draw may indicate a fuel flow restriction, an electric supply concern or a worn pump.

☐ **2.** Check the service information for the fuel pump current draw test procedures.

☐ **3.** Connect the digital multimeter, set to read amperes (A) and connect the red lead to the positive (+) of the battery. Connect the black lead to the fuel pump test terminal. The pump should run, and an amperage reading should be observed on the meter. (Allow the pump to run for 30 seconds.) Confirm the reading with acceptable specifications.

Reading = ☐ amp

Normal readings: 2 to 5 amps (9-13 psi)
4 to 8 amps (35-45 psi)
8 to 12 amps (55-64 psi)

☐ **4.** Based on the test results, what is the needed action?

Pearson

Education Foundation

Fuel Filter Replacement

Meets ASE Task: A8 – D-2 – P-2

Evaluation (Enter number from 4, 3, 2, 1): ☐

Name: ☐ **Date:** ☐ **Time on Task:** ☐

Make/Model/Year: ☐ **VIN:** ☐

☐ **1.** Check service information for the recommended part number and fuel filter replacement procedure.

a. Fuel filter part number = ☐

b. Recommended procedure: ☐

☐ **2.** Check all that apply:

☐ ☐ Fuel system pressure should be relieved before removing the old filter.

☐ ☐ Special tools required to remove filter (describe): ☐

☐ ☐ Direction of fuel flow labeled on the filter. ☐

☐ ☐ Other (describe) ☐

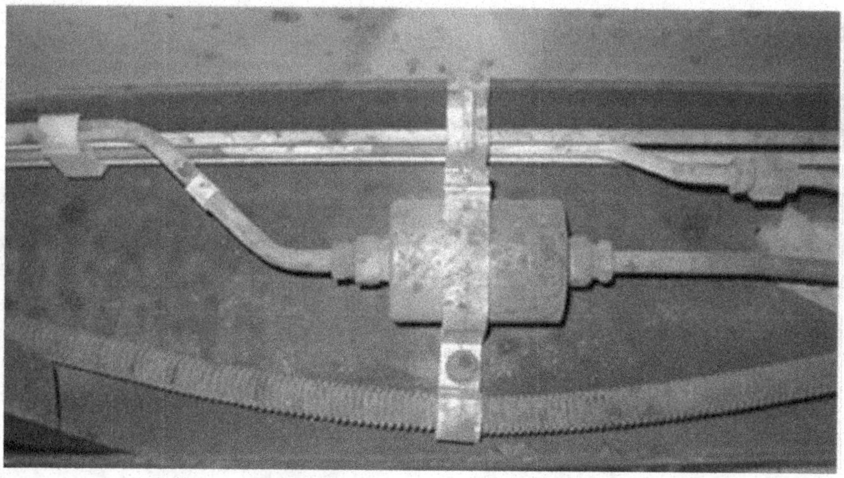

Pearson

Education Foundation

Modes of Fuel Injection Operation

Meets ASE Task: A8 – A-1 – P-1

Evaluation (Enter number from 4, 3, 2, 1): ☐

Name: ☐ **Date:** ☐ **Time on Task:** ☐

Make/Model/Year: ☐ **VIN:** ☐

☐ **1.** Check service information for the sensors that are used to control the fuel injection system at each of the five modes of operation.

☐ **2.** Starting cranking mode:

Sensors Used	Why

☐ **3.** Acceleration enrichment mode:

Sensors Used	Why

☐ **4.** Deceleration enleanment mode:

Sensors Used	Why

☐ **5.** Open-loop operation mode:

Sensors Used	Why

☐ **6.** Closed-loop operation mode:

Sensors Used	Why

Pearson

Education Foundation

Idle Air Control

Meets ASE Task: A8 – D-9 – P-1

Evaluation (Enter number from 4, 3, 2, 1): ☐

Name: ☐ **Date:** ☐ **Time on Task:** ☐

Make/Model/Year: ☐ **VIN:** ☐

☐ **1.** The idle air control is used to control idle speed by increasing or decreasing the amount of air entering the engine like what occurs when the accelerator pedal is depressed.

☐ **2.** Connect a scan tool.

☐ **3.** Look at the IAC commanded position = ☐ (should be 15 to 25% or counts on a warm engine in park or neutral).

OK ☐ ☐ NOT OK ☐ ☐

☐ **4.** Diagnosis:

IAC counts higher than normal. This could indicate one or more of the following:

1. Engine not fully warm
2. Some electrical load is on, such as daytime running lights or air conditioning
3. Dirty throttle plates
4. Abnormal load on the engine

IAC counts lower than normal. This could indicate one or more of the following:

1. A vacuum leak
2. Misadjusted idle speed control
3. Stuck or binding throttle cable or linkage

☐ **5.** Based on the inspection of the system, what is the necessary action?

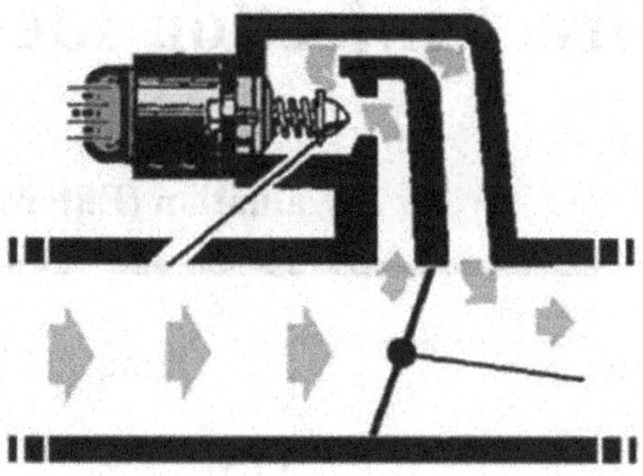

Pearson

Education Foundation

Gasoline Direct Injection Identification

Meets ASE Task: A8 – A-1 – P-1

Evaluation (Enter number from 4, 3, 2, 1): ☐

Name: ☐ **Date:** ☐ **Time on Task:** ☐

Make/Model/Year: ☐ **VIN:** ☐

☐ **1.** Check service information to determine if the vehicle being serviced is equipped with gasoline direct injection. (Check all that apply)

☐ Gasoline direct injection only

☐ Port and gasoline direct injection system

☐ **2.** List the specifications as found in service information.

a. Lift pump pressure = ☐

b. Lift pump volume = ☐

c. High-pressure system pressure = ☐

d. Fuel injector resistance = ☐

e. Other (describe) ☐

☐ **3.** What are the specified maintenance procedures required (if any)? Describe:

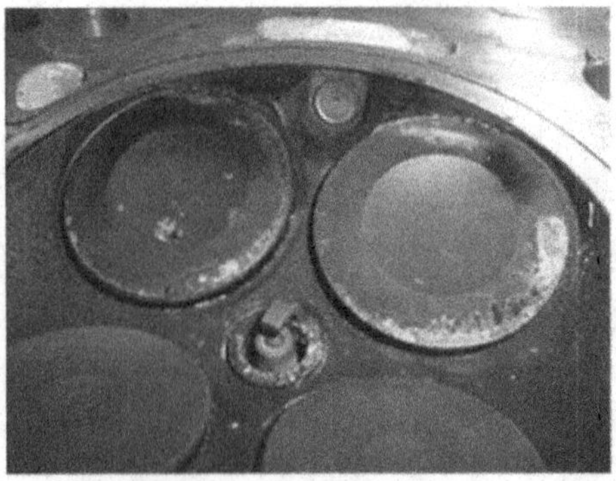

Pearson

ASE Education Foundation

Scan Tool Diagnosis

Meets ASE Task: A8 – A-2 – P-1 & A8 – B-5 – P-1

Evaluation (Enter number from 4, 3, 2, 1): ☐

Name: ☐ **Date:** ☐ **Time on Task:** ☐

Make/Model/Year: ☐ **VIN:** ☐

☐ **1.** DTCs: ☐ Pending DTC: ☐

☐ **2.** Engine coolant temperature (ECT): cold ☐ warm ☐

☐ **3.** Intake air temperature (IAT): cold ☐ warm ☐

☐ **4.** Upstream O2S1: lowest voltage observed ☐ highest voltage observed ☐

☐ **5.** Upstream O2S2: lowest voltage observed ☐ highest voltage observed ☐

☐ **6.** Downstream O2S: lowest voltage observed ☐ highest voltage observed ☐

☐ **7.** Injector pulse width: @ idle (park) ☐ @ 2000 RPM (park) ☐

☐ **8.** Spark advance: @ idle (park) ☐ @ 2000 RPM (park) ☐

☐ **9.** Short term fuel trim: @ idle (park) ☐ @2000 RPM (Park) ☐

☐ **10.** Long term fuel trim (block learn): @ idle (park) ☐ @ idle (drive) ☐

☐ **11.** MAP: @ idle (park) ☐ @ idle (drive) ☐

☐ **12.** MAF (grams/sec): @ idle (park) ☐ @ 2000 RPM (park) ☐

☐ **13.** IAC counts: @ idle (park) ☐ @ idle (drive) ☐

☐ **14.** Throttle position (TP) sensor: @ idle ☐ @ W.O.T. engine off, ignition "on" ☐

☐ **15.** Battery voltage: ☐

(Not all of this data will be available on all vehicles or scan tools - if not available, put NA.)

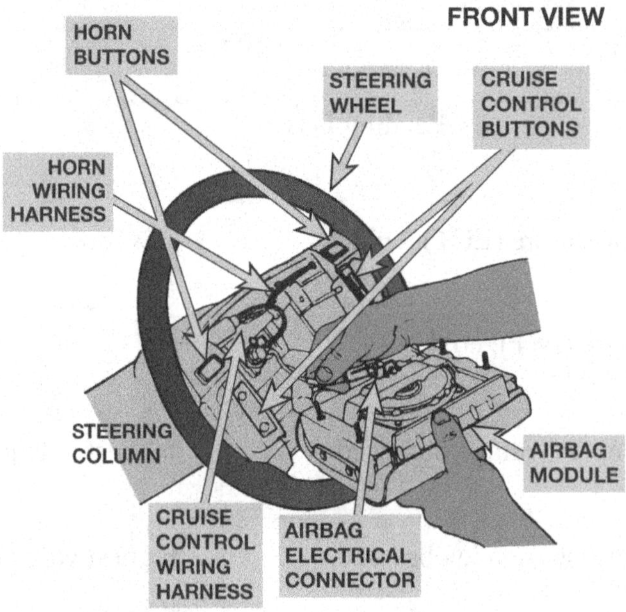

Pearson

Education Foundation

Fuel Trim Diagnosis II

Meets ASE Task: A8 – B-5 – P-1

Evaluation (Enter number from 4, 3, 2, 1): ☐

Name: ☐ **Date:** ☐ **Time on Task:** ☐

Make/Model/Year: ☐ **VIN:** ☐

☐ **1.** Connect a scan tool and select long-term fuel trim (LTFT) and short-term fuel trim (STFT).

☐ **2.** Start the engine and operate until normal operating temperature and closed loop status is achieved.

☐ **3.** Record the following fuel trim information:

	Cell #	LTFT	STFT
Idle in Drive (if automatic transmission only)			
Idle in Park A/C off			
Idle in Park A/C on			
3000 RPM in Park			

☐ **4.** Based on the test results, what is the needed action?

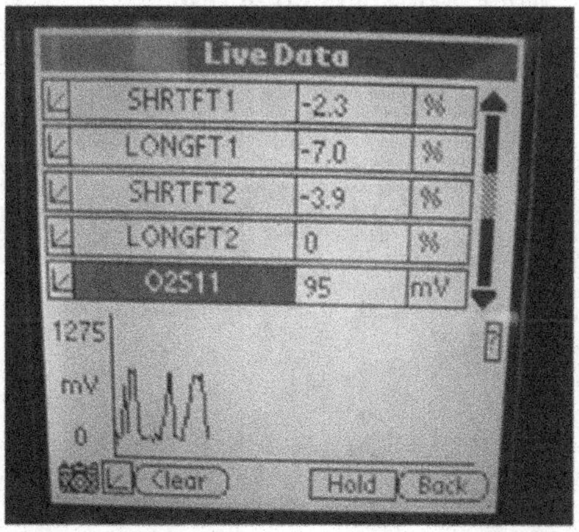

Fuel-Injection System Diagnosis

Meets ASE Task: A8 – D-13 – P-2

Evaluation (Enter number from 4, 3, 2, 1): ☐

Name: ☐ **Date:** ☐ **Time on Task:** ☐

Make/Model/Year: ☐ **VIN:** ☐

☐ **1.** Check service information for the recommended procedure to follow to diagnose the fuel injection system.

☐ **2.** Attach a fuel pressure gauge to the Schrader valve on the fuel rail, if available.

☐ **3.** Turn the ignition key to "on" or start the engine to build up the fuel pump pressure. Record the fuel pump pressure. ☐ psi

☐ **4.** Turn the ignition off and wait 20 minutes and observe the fuel pressure retained in the fuel rail = ☐ psi.

If the drop is greater than 20 psi in 20 minutes, there is a possible problem with:

a. the check valve in the fuel pump.
b. leaking injectors.
c. a defective (leaking) fuel pressure regulator.

☐ **5.** Based on the test results, what is the needed action?

Pearson

Education Foundation

Injector Resistance Testing

Meets ASE Task: A8 – D-10 – P-1

Evaluation (Enter number from 4, 3, 2, 1): ☐

Name: ☐ **Date:** ☐ **Time on Task:** ☐

Make/Model/Year: ☐ **VIN:** ☐

☐ **1.** Check service information for the specified injector resistance

☐ **2.** Measure the resistance of all injectors with a digital ohmmeter.

Injector resistance:

	Resistance		Resistance
1.		5.	
2.		6.	
3.		7.	
4.		8.	

NOTE: For best performance and idle quality, all injectors should measure within 0.3 to 0.4 ohms of each other.

☐ **3.** Based on the resistance test results, what action is needed?

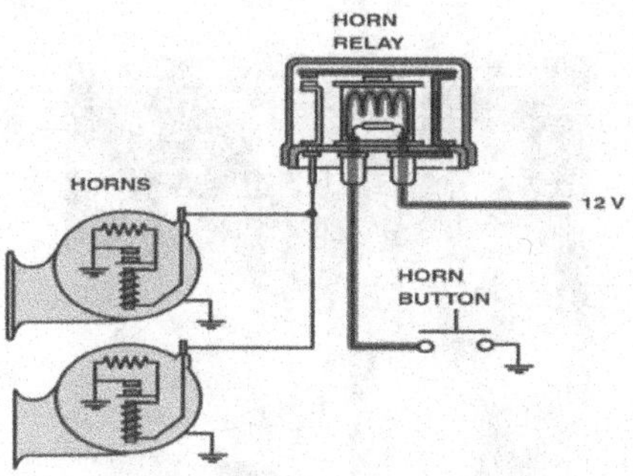

Pearson

Education Foundation

Fuel Injector Balance Test

Meets ASE Task: A8 – D-10 – P-1

Evaluation (Enter number from 4, 3, 2, 1): ☐

Name: ☐ **Date:** ☐ **Time on Task:** ☐

Make/Model/Year: ☐ **VIN:** ☐

☐ **1.** Check service information for the procedures and specifications for an injector voltage drop test.

☐ **2.** Record the voltage drop of the injectors (pressure after injector was pulsed on).

Cylinder #1		Cylinder #5	
Cylinder #2		Cylinder #6	
Cylinder #3		Cylinder #7	
Cylinder #4		Cylinder #8	

☐ **3.** What is the maximum allowable variation in voltage drop among the injectors?

☐ **4.** Compare the test results with the specifications. What is the needed action?

Pearson

Education Foundation

Injector Voltage Waveform Test

Meets ASE Task: A8 – D-10 – P-1

Evaluation (Enter number from 4, 3, 2, 1): ☐

Name: ☐ **Date:** ☐ **Time on Task:** ☐

Make/Model/Year: ☐ **VIN:** ☐

☐ **1.** Check service information for the type of fuel injector being used.

- ☐ Saturated
- ☐ Peak and hold

☐ **2.** Connect a digital storage oscilloscope (DSO) or graphing multimeter (GMM) to the pulsed side of the injector. (Check service information for the color of wire used for the pulse.)

☐ **3.** Start the engine and observe the voltage waveform.

☐ **4.** Does the voltage spike (kick) exceed 50 volts? ☐ Yes ☐ No

☐ **5.** What is the injector pulse-width? ☐ (normally between 1.5 and 3.5 MS at idle on a warm engine)

☐ **6.** Based on the test performed, what is the needed action?

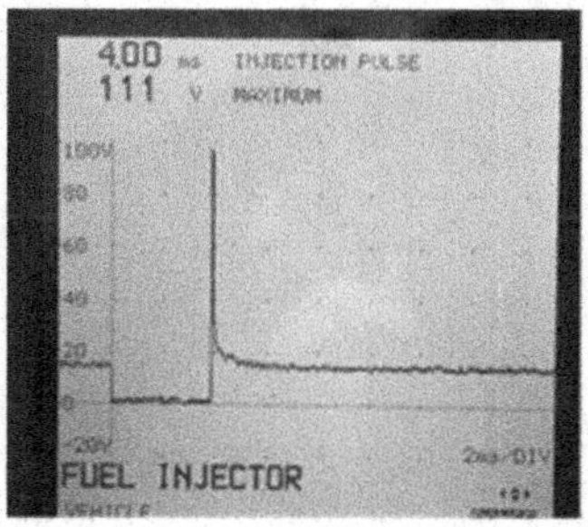

Pearson

Education Foundation

Module Communication II

Meets ASE Task: A6 - G-8 – P-1

Evaluation (Enter number from 4, 3, 2, 1): [ ]

Name: [ ] **Date:** [ ] **Time on Task:** [ ]

Make/Model/Year: [ ] **VIN:** [ ]

☐ **1.** Check service information for the specified method and procedures to follow to check for proper module communications.

☐ **2.** Check all that apply:

- ☐ [ ] Use a factory scan tool.
- ☐ [ ] Use a generic OBD-II scan tool.
- ☐ [ ] Use a DMM to check for resistance on communication circuits.
- ☐ [ ] Use a fused jumper wire and a scan tool to diagnose communication errors.
- ☐ [ ] Other (describe) [ ]

☐ **3.** List the modules that are present in the vehicle being tested.

a.	f.
b.	g.
c.	h.
d.	i.
e.	j.

☐ **4.** Based on the test results, what is the necessary action?

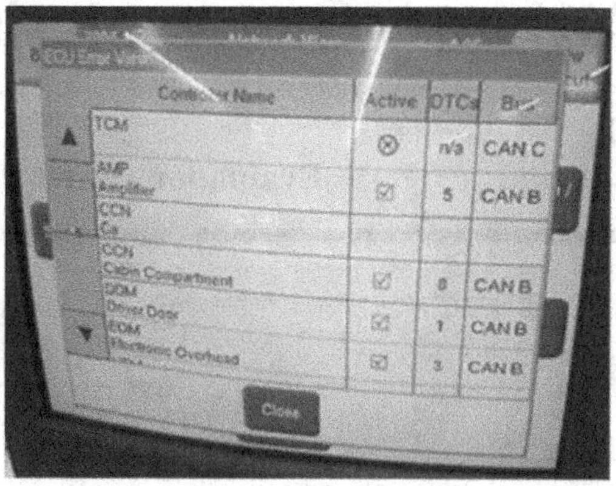

Pearson

Education Foundation

Exhaust Gas Analysis

Meets ASE Task: A8 – E-1 – P-1

Evaluation (Enter number from 4, 3, 2, 1): ☐

Name: ☐ **Date:** ☐ **Time on Task:** ☐

Make/Model/Year: ☐ **VIN:** ☐

☐ **1.** Check the instruction information for the exhaust gas analyzer being used to determine the proper test procedures to follow.

☐ **2.** Check the vehicle for exhaust leaks and other faults that could affect the exhaust gas readings.

☐ **3.** Prepare the vehicle for testing, which usually includes operating the engine until normal operating temperature has been achieved. List other items listed by the test equipment manufacturer that should be performed.

☐ **4.** Obtain the exhaust gas readings and compare them to specifications.

Gas	Idle	2500 RPM	General Specifications
HC			Max 50 PPM
CO			Max 0.5%
CO_2			12% to 15% or higher
O_2			0% to 2%
NO_X			Less than 100 PPM @ idle Less than 1000 PPM @ wide open throttle

☐ **5.** Based on the exhaust gas readings, what is the needed action?

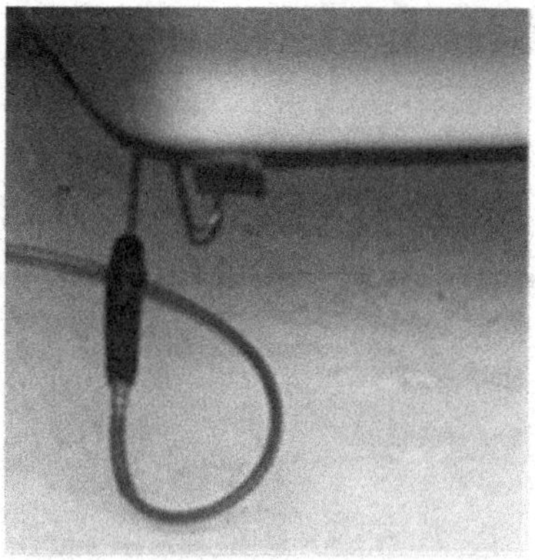

Pearson

Education Foundation

EVAP System Component Inspection

Meets ASE Task: A8 – E-7 – P-1

Evaluation (Enter number from 4, 3, 2, 1): ☐

Name: ☐ **Date:** ☐ **Time on Task:** ☐

Make/Model/Year: ☐ **VIN:** ☐

☐ **1.** According to the under-hood emission label, with what type of system is the vehicle equipped?

- ☐ Pre-OBD I (1987 or older)
- ☐ OBD I (1988-1995)
- ☐ OBD II (1996 and newer)
- ☐ Unknown (describe): ☐

☐ **2.** Check the service information and the vehicle emission label for the following information.

a. The location of the carbon canister (describe) ☐

b. The location of the purge control solenoid(s) (describe) ☐

c. The resistance specification for the evaporator control solenoid(s) = ☐

☐ **3.** Visually inspect the following items and check for any sign of a leak in the system.

a. Canister hoses OK ☐ ☐ NOT OK ☐ ☐

b. Gas cap OK ☐ ☐ NOT OK ☐ ☐

☐ **4.** Based on the inspection, what is the needed action?

Pearson

Education Foundation

Evaporative Emission Controls Diagnosis

Meets ASE Task: A8 – E-7 – P-1

Evaluation (Enter number from 4, 3, 2, 1): ☐

Name: ☐ **Date:** ☐ **Time on Task:** ☐

Make/Model/Year: ☐ **VIN:** ☐

☐ **1.** Check service information for the specified tests and procedures to follow to diagnose the problems in the evaporative emission control system.

☐ **2.** List the tools and equipment specified for use by service information. Check all that apply.

- ☐ ☐ Special tester (describe) ☐
- ☐ ☐ Scan tool
- ☐ ☐ Other (describe) ☐

☐ **3.** List the components included in the evaporative emission control unit and describe how each is to be tested according to service information.

	Component	Test or Inspection
a.		
b.		
c.		
d.		

☐ **4.** Based on the results of the tests and inspection, what is the needed action?

Pearson

Education Foundation

Evaporative Emission DTC Diagnosis

Meets ASE Task: A8 – E-8 – P-1

Evaluation (Enter number from 4, 3, 2, 1): ☐

Name: ☐ **Date:** ☐ **Time on Task:** ☐

Make/Model/Year: ☐ **VIN:** ☐

☐ **1.** Check service information for the specified diagnostic steps to follow if evaporative emission control-related DTC(s) is set.

☐ **2.** Describe the test and inspection specified for each of the DTCs listed.

DTC	Specified Test(s)
P0440	
P0441	
P0442	
Other (list)	
Other (list)	

☐ **3.** Based on the results of the tests and inspection, what is the needed action?

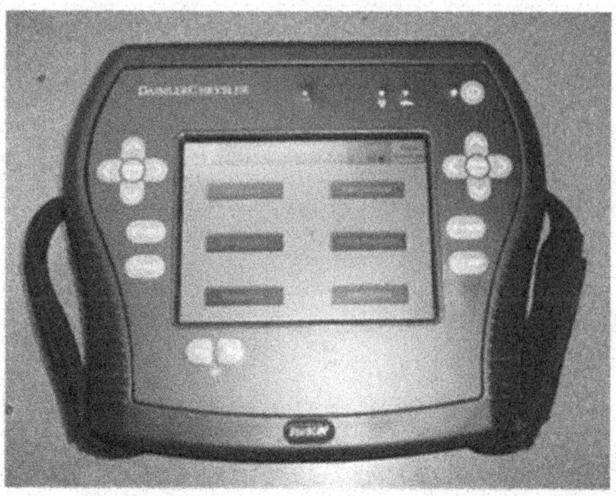

Pearson

Education Foundation

EGR System Diagnosis

Meets ASE Task: A8 – E-4 – P-1

Evaluation (Enter number from 4, 3, 2, 1): ☐

Name: ☐ **Date:** ☐ **Time on Task:** ☐

Make/Model/Year: ☐ **VIN:** ☐

☐ **1.** Check service information for the specified testing procedure of the exhaust gas recirculation (EGR) system using a scan tool.

☐ **2.** List the EGR-related data that can be retrieved using a scan tool.

☐ **3.** List the scan tool commands (bi-directional) for the EGR system and describe the results of the tests.

Unit Commanded	Results

☐ **4.** Has the EGR OBD II monitor run? ☐

☐ **5.** Based on the results of the scan tool diagnosis, what is the needed action?

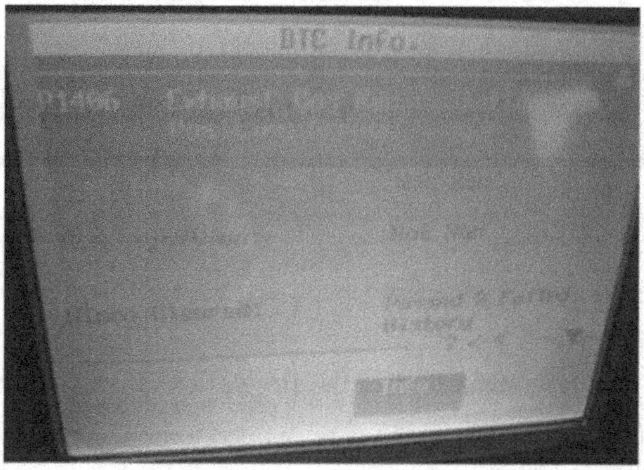

Pearson

Education Foundation

Service EGR System

Meets ASE Task: A8 – E-4 – P-1

Evaluation (Enter number from 4, 3, 2, 1): ☐

Name: ☐ **Date:** ☐ **Time on Task:** ☐

Make/Model/Year: ☐ **VIN:** ☐

☐ **1.** The EGR passages and valve that control the flow of exhaust gases can become clogged with carbon. The EGR valve and passages may need to be cleaned if one or more of the following conditions are present.

- A computer diagnosis trouble code (DTC) indicating the lack of EGR flow.
- The failure of an exhaust emission test for excessive NOx
- Excessive engine spark knock (ping or detonation)

☐ **2.** Check service information for the recommended procedures to follow when servicing the EGR system.

☐ **3.** Remove the EGR valve and inspect for clogged passages. Clean as needed.

Valve was clogged ☐ Valve was OK ☐

☐ **4.** To clean the passages of carbon, remove the plugs or EGR valve and insert a stiff wire into an electric drill and use it to ream out the passages.

☐ **5.** Reinstall the EGR valve with a new gasket and check the engine for proper operation.

Pearson

Education Foundation

EGR Electrical Sensors

Meets ASE Task: A8 – E-4 – P-1

Evaluation (Enter number from 4, 3, 2, 1): ☐

Name: ☐ **Date:** ☐ **Time on Task:** ☐

Make/Model/Year: ☐ **VIN:** ☐

☐ **1.** Check service information for the recommended tests and diagnostic procedure to follow to diagnose EGR system sensors and controls.

☐ **2.** List the tools and equipment needed as specified by the vehicle manufacturer. Check all that apply.

- ☐ ☐ Scan tool
- ☐ ☐ Vacuum pump
- ☐ ☐ Digital multimeter (DMM)
- ☐ ☐ 5-gas exhaust analyzer
- ☐ ☐ Other (describe) ☐

☐ **3.** Based on the test results, what is the needed action?

Pearson

Education Foundation

PCV System Inspection

Meets ASE Task: A8 – E-2 – P-2

Evaluation (Enter number from 4, 3, 2, 1): ☐

Name: ☐ **Date:** ☐ **Time on Task:** ☐

Make/Model/Year: ☐ **VIN:** ☐

☐ **1.** Check service information for the recommended steps to follow when testing or servicing the positive crankcase ventilation (PCV) system.

☐ **2.** Check service information and describe the location of the following:

PCV valve ☐

Crankcase vent filter ☐

Fixed orifice (if equipped) ☐

Other (describe) ☐

☐ **3.** What is specified replacement interval for the PCV valve?

☐ **4.** Remove and clear the PCV valve (if equipped) and note the condition.

☐ Like new ☐ Very dirty

☐ Slightly dirty ☐ Valve clogged or stuck.

☐ Other (describe) ☐

☐ **5.** Based on the test and inspection and on the recommendation of the vehicle manufacturer, what is the needed action?

Pearson

Education Foundation

PCV System Diagnosis

Meets ASE Task: A8 – E-2 – P-2

Evaluation (Enter number from 4, 3, 2, 1): ☐

Name: ☐ **Date:** ☐ **Time on Task:** ☐

Make/Model/Year: ☐ **VIN:** ☐

☐ **1.** Check service information for the recommended procedures to follow when diagnosing the PCV system.

☐ **2.** Start the engine and allow it to idle and remove the oil fill cap.

☐ **3.** Place a piece of paper or a 3" x 5" card over the filler. (The PCV system is functioning correctly if the paper is held down tight onto the filler by vacuum in the crankcase).

☐ **4.** Seal off the oil fill opening and measure the crankcase vacuum at the dipstick tube = ☐ (should be about 0.5 in. Hg. or 7 in. or more of water if using a water manometer).

OK ☐ ☐ NOT OK ☐ ☐

☐ **5.** Based on the test results, what is the needed action?

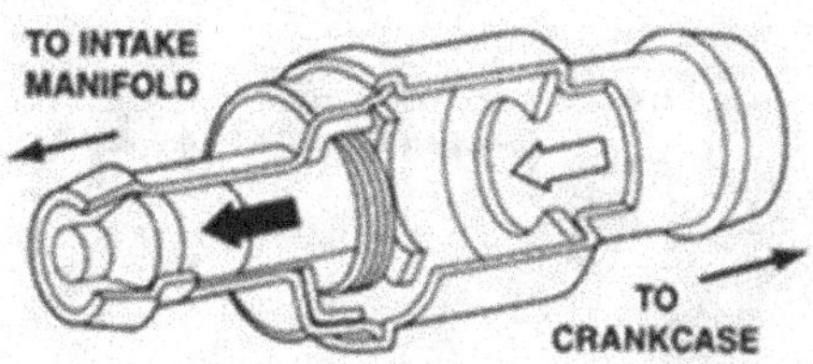

Pearson

Education Foundation

Secondary Air Injection Diagnosis

Meets ASE Task: A8 – E-5 – P-3

Evaluation (Enter number from 4, 3, 2, 1): ☐

Name: ☐ **Date:** ☐ **Time on Task:** ☐

Make/Model/Year: ☐ **VIN:** ☐

☐ **1.** Check service information for the recommended test procedure and specifications for the secondary air injection system.

☐ **2.** Locate the air pump.

☐ **3.** Carefully inspect the condition of all the hoses, check the valves and the metal lines for corrosion or damage.

☐ **4.** Start the engine and feel the air pump lines to confirm the proper air flow.

☐ **5.** Inspect the air pump drive belt for cracks and proper tension or electrical connections for an electric air pump.

☐ **6.** Based on the inspection and test results, what is the needed action?

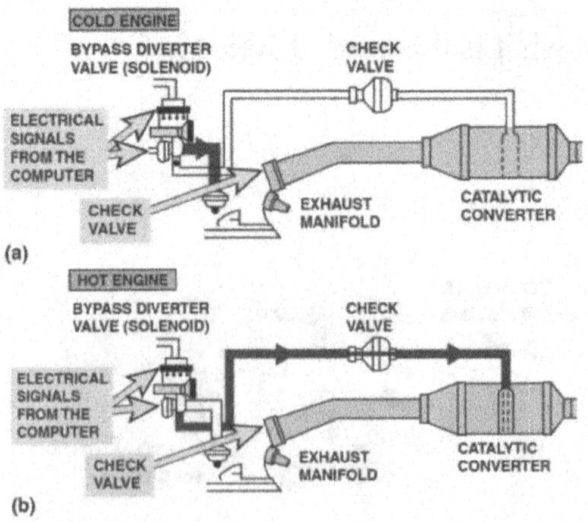

Pearson

ASE Education Foundation

AIR Pump Component Inspection

Meets ASE Task: A8 – E-5 – P-3

Evaluation (Enter number from 4, 3, 2, 1):

Name: **Date:** **Time on Task:**

Make/Model/Year: **VIN:**

☐ **1.** Check service information for the recommended procedures to follow when inspecting and testing the AIR pump components.

☐ **2.** Carefully inspect the condition of all the hoses, check the valves and the metal lines for corrosion or damage.

☐ **3.** Start the engine and feel the air pump lines to confirm the proper air flow.

☐ **4.** Inspect the air pump drive belt for cracks and proper tension.

☐ **5.** Based on the inspection, what is the needed action?

NOTE: A defective one-way check valve at the exhaust manifold can allow hot exhaust gases to flow past the check valve and cause damage to the switching valves, hoses, or air pump itself. These exhaust gases can cause poor engine operation and stalling if drawn into the air intake system.

Pearson

Education Foundation

AIR Pump Electrical Component Inspection

Meets ASE Task: A8 – E-5 – P-3

Evaluation (Enter number from 4, 3, 2, 1): ☐

Name: ______ **Date:** ______ **Time on Task:** ______

Make/Model/Year: ______ **VIN:** ______

☐ **1.** Check service information for the recommended procedures, tests, and specifications for inspecting and testing the electrical/electronic component of the secondary air injection system.

☐ **2.** List the tools and equipment specified by the vehicle manufacturer to diagnose the electrical/electronic components of the secondary air injection system. Check all that apply.

- ☐ ☐ Digital multimeter (DMM)
- ☐ ☐ Scan tool
- ☐ ☐ Exhaust gas analyzer
- ☐ ☐ Other (describe) ______

☐ **3.** Based on the inspection and tests, what is the needed action?

Pearson

Education Foundation

Exhaust System Back Pressure Test

Meets ASE Task: Task details here

Evaluation (Enter number from 4, 3, 2, 1): ☐

Name: ______ **Date:** ______ **Time on Task:** ______

Make/Model/Year: ______ **VIN:** ______

☐ **1.** Check service information for the specified maximum back pressure (record below).

☐ **2.** Remove the oxygen sensor from the exhaust manifold and install a tool to measure exhaust back pressure.

☐ **3.** Connect a vacuum/pressure gauge to the exhaust back pressure tool. Start the engine and run at idle and observe exhaust back pressure.

______ psi back pressure

☐ **4.** Operate the engine at a constant speed of 2500 RPM and observe the exhaust back pressure.

______ psi back pressure

☐ **5.** Based on the results of the backpressure test, what is the necessary action?

Pearson

Education Foundation

Catalytic Converter Performance Test

Meets ASE Task: Task details here

Evaluation (Enter number from 4, 3, 2, 1): ☐

Name: ☐ **Date:** ☐ **Time on Task:** ☐

Make/Model/Year: ☐ **VIN:** ☐

☐ **1.** Start the engine and run at a fast idle (2500 RPM) for at least 2 minutes to fully warm up the oxygen sensor, the engine coolant and catalytic converter.

☐ **2.** Using a pyrometer (infrared or contact type), measure the front (inlet) and outlet of the catalytic converter.

Inlet temperature = ☐°

Outlet temperature = ☐°

Difference = ☐°

☐ **3.** Results: If the outlet temperature is 50°F (10°C) (or 10%) higher than the inlet temperature, the catalytic converter is functioning correctly. OK ☐ ☐ NOT OK ☐ ☐

☐ **4.** Based on the test results, what is the necessary action?

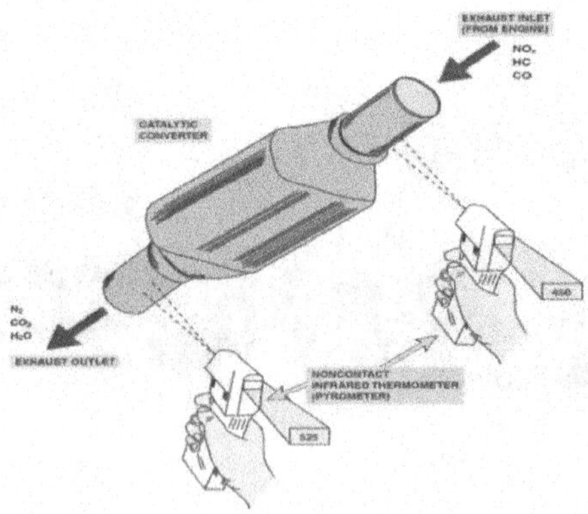

Pearson

Education Foundation

Turbocharger/ Supercharger Diagnosis

Meets ASE Task: Task details here

Evaluation (Enter number from 4, 3, 2, 1): ☐

Name: ☐ **Date:** ☐ **Time on Task:** ☐

Make/Model/Year: ☐ **VIN:** ☐

☐ **1.** Check service information for the recommended procedure and specifications to follow to test the operation of the turbocharger/supercharger system.

☐ **2.** Is the turbocharged/supercharged vehicle equipped with an intercooler?

- ☐ ☐ Yes (If so, describe the type.) ☐
- ☐ ☐ No

☐ **3.** What tools and equipment are needed to test the operation of the turbocharger/supercharger system? Check all that apply.

- ☐ ☐ Scan tool
- ☐ ☐ Vacuum/pressure gauge
- ☐ ☐ Infrared pyrometer
- ☐ ☐ Other (describe) ☐

☐ **4.** Based on the diagnosis of the turbocharger/supercharger system, what is the necessary action?

Pearson

ASE Education Foundation

Exhaust System Inspection

Meets ASE Task: Task details here

Evaluation (Enter number from 4, 3, 2, 1): ☐

Name: ☐ **Date:** ☐ **Time on Task:** ☐

Make/Model/Year: ☐ **VIN:** ☐

☐ **1.** Safely hoist the vehicle and wear safety glasses.

☐ **2.** Visually inspect the following items and note their condition.

Tailpipe OK ☐ NOT OK ☐

Describe fault: ☐

Muffler OK ☐ NOT OK ☐

Describe fault: ☐

Exhaust Pipe OK ☐ NOT OK ☐

Describe fault: ☐

Catalytic converter OK ☐ NOT OK ☐

Describe fault: ☐

Exhaust manifold(s) OK ☐ NOT OK ☐

Describe fault: ☐

Hangers & Heat shields(s) OK ☐ NOT OK ☐

Describe fault: ☐

☐ **3.** Based on the inspection, what is the necessary action?

Pearson

Education Foundation

Inspect and Replace Air Filter

Meets ASE Task: Task details here

Evaluation (Enter number from 4, 3, 2, 1): ☐

Name: ☐ **Date:** ☐ **Time on Task:** ☐

Make/Model/Year: ☐ **VIN:** ☐

☐ **1.** Check service information for the specified procedure to follow to inspect and replace an air filter. Describe the specified procedure.

☐ **2.** Check filter housing and intake duct work.

- ☐ ☐ OK
- ☐ ☐ NOT OK (describe) ☐

☐ **3.** The air filter part number is ☐

Pearson

Education Foundation

Diesel Exhaust Fluid

Meets ASE Task: Task details here

Evaluation (Enter number from 4, 3, 2, 1): ☐

Name: ☐ **Date:** ☐ **Time on Task:** ☐

Make/Model/Year: ☐ **VIN:** ☐

☐ **1.** Check service information for the location of the diesel exhaust fluid fill port.

Describe the location. ☐

☐ **2.** Check service information and determine the capacity of the DEF storage container.

DEF storage container capacity = ☐

☐ **3.** Check the DEF level and refill to specifications (record your findings).

Pearson

Education Foundation

Module Programming III

Meets ASE Task: A8 – C-5 – P-2

Evaluation (Enter number from 4, 3, 2, 1): ☐

Name: ☐ **Date:** ☐ **Time on Task:** ☐

Make/Model/Year: ☐ **VIN:** ☐

☐ **1.** Describe the condition(s) that requires the module to be reprogrammed.

☐ **2.** What J2534 programming hardware is being used? (Describe).

☐ **3.** What battery maintainer (charger) was used to keep the vehicle battery charged during the reprogramming? List the make and model:

☐ **4.** Describe the results of the reprogramming.

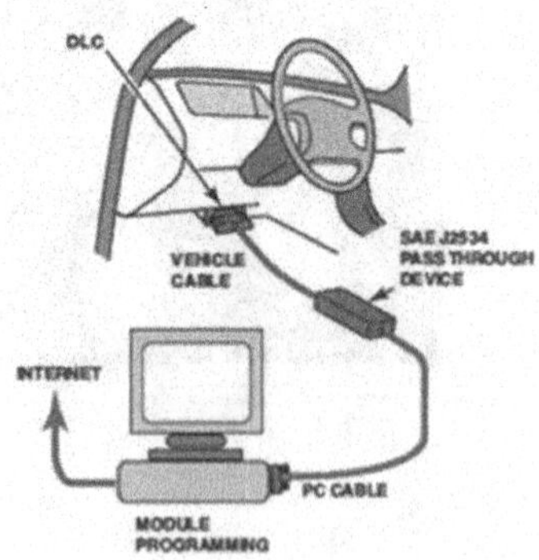